T0260353

The American
Phage Group

The American Phage Group

Founders of Molecular Biology

WILLIAM C. SUMMERS

Yale UNIVERSITY PRESS

New Haven and London

Published with assistance from the foundation established in memory of James Wesley Cooper of the Class of 1865, Yale College, and from the income of the Frederick John Kingsbury Memorial Fund.

Yale University Press books may be purchased in quantity for educational, business, or promotional use. For information, please e-mail sales.press@yale.edu (U.S. office) or sales@yaleup.co.uk (U.K. office).

Set in Minion type by Integrated Publishing Solutions
Printed in the United States of America.

ISBN 978-0-300-26356-5 (hardcover : alk. paper)
Library of Congress Control Number: 2022932737
A catalogue record for this book is available from the British Library.

This paper meets the requirements of ANSI/NISO z39.48-1992 (Permanence of Paper).

10 9 8 7 6 5 4 3 2 1

For Wacław Szybalski (1921–2020),
Scientist, Teacher, Friend

. . . complete ignorance and infinite intelligence.

—MAX DELBRÜCK

Contents

Preface

Molecular biology is now recognized as a distinct approach to the study of the living natural world, an approach that differs from classical biology in several ways: its units of analysis, molecules rather than organisms; its methodological approaches, laboratory interventions rather than field observations; and the types of answers sought, based on physical and chemical principles rather than morphological and behavioral generalizations. In the middle decades of the twentieth century molecular biology evolved into a recognizable discipline. How did this new discipline arise? From the foundations of classical biology or from an entirely new coalescence of ideas and approaches from outside classical biology?

Historians of science have pursued such questions with vigor. A simple Google Scholar search for the phrase "origins of molecular biology" yielded over 2,600 citations. One recurrent theme in this historiography has been the role of bacteriophages, viruses that infect bacteria, in the history of molecular biology. For a small group of scientists between 1940 and 1960, these tiny objects were the focus of intense study as basic models for the gene and for the process of reproduction. Some of these bacteriophage enthusiasts came to be called the American Phage Group. They came from diverse backgrounds, not primarily classical biology, bringing new viewpoints, methods, experimental standards, and laboratory cultures. The research framework developed by this group of bacteriophage workers, along with complementary influences from structural biology, provided the foundation upon which molecular biology was built.

The history of the American Phage Group has been told in other versions: recollections of the participants, biographical studies of celebrated individuals, and narratives by historians of science. This book goes beyond those accounts in several ways. It recognizes the complexities of the American Phage Group in new ways. This book does not set forth a linear narrative of progress by past heroes; instead it emphasizes the diversity and historical contingencies in the group's development. And, to avoid post hoc story construction, this account is richly grounded in archival sources and contemporary records.

To illuminate how the American Phage Group formed, grew, and influenced the new discipline of molecular biology, this account employs a conceptual model for discipline formation based on "research framework theory" developed by Barbara Von Eckardt in her study of the origins of cognitive science. This is not a "strong theory," but it seems an adaptable and useful way to organize complex stories of nascent discipline formation.

In the reconstruction that follows, birth and death dates (where known) will be provided at the first mention of a scientist's name, to help the reader better understand the chronology. Some of the interviews with subjects of this account were conducted informally, without verbatim recordings but with extensive notes; I have limited direct quotations to the comments transcribed in those notes.

Acknowledgments

First and foremost, I am indebted to my late colleague and friend Frederic L. (Larry) Holmes for his inspiration, guidance, and support over many years. Other colleagues who have contributed much to my education as a historian of science include Daniel Kevles, Pnina Abir-Am, Neeraja Sankaran, Nathaniel Comfort, and the community of scholars associated with the Joint Atlantic Seminar on the History of Biology. Essential help has been provided by archivists and librarians at the Pasteur Institute, the California Institute of Technology, the Bancroft Library at the University of California, Berkeley, and, especially, at the Cold Spring Harbor Laboratory. My scientific colleagues have been generous with their time and recollections, especially Wacław Szybalski, my doctoral adviser, to whom this work is dedicated, but also Seymour Benzer, Joe Bertani, Michael Cohen, Seymour Cohen, Jonathan Delbrück, Emory Ellis, Bert Hansen, Lloyd Kozloff, Bernard Latarjet, Raymond Latarjet, Joshua Lederberg, Bill Pirie, Peggy Lieb, Nicolas Rasmussen, Jim Strick, Jan Witkowski, and Elie Wollman. Numerous other phage biologists and phage course alumni responded to my inquiries with valuable insights. Barbara Von Eckardt's book on discipline formation provided the initial organizing framework for this work. As always, I am grateful to Wilma P. Summers for more than she knows.

1

Life, Genes, and Phages

A Campfire Story (circa 1965)

In the beginning there were the Great Physicists—Planck, Einstein, and Bohr—and they created the world. Then came the Good Physicists, but soon there was little work left for them to do, so they searched for new games to play and they discovered biology. Biology was still a dismal swamp of myth and superstition, ruled by the cult of the morphologists, the cabal of the embryologists, and the slightly more enlightened geneticists. "Aha!" said the Good Physicists: "We have solved the riddle of the atom; now we will solve the riddle of the cell," and so they did, leaving the biologists to wither away in irrelevance or to become new Good Physicists, or molecular biologists, as they now called themselves. The leader of the Good Physicists was called Delbrück and his disciples were called "the Phage Group."

This was the view of the origin of molecular biology that I learned by heart along with an exegetical tradition known as "Delbrück Stories" that circulated when I was trying to become a molecular biologist in the mid-1960s. Molecular biology is now an accepted genre of the biological sciences. Whether it warrants recognition as a distinct discipline may be debated, but with academic departments, journals, textbooks, and many scientists adopting the designation, it seems reasonable or at least heuristic to consider "Molecular Biology" as an established kind of biology with a stable position in modern science. Just how this field, discipline, or way of thinking originated, evolved, and matured is the subject of this book. This process of what I call "discipline formation" took place (roughly) between the 1930s and the 1950s. Molecular biology is usually taken to stand in distinction to classical biology, organismal biology, or general biology, all of which were heavily dependent on generalizations from morphological studies and field behavioral research. What distinguished molecular biology was its exploration of chemical and physical explanations and experimentation as the central focus of investigation.

Two main research pathways that merged in the early 1950s led, in a very general way, to the formation of the discipline of molecular biology. One research program sought a detailed materialistic explanation of the mechanism of heredity— "how does like beget like?"—while the other research program sought to understand how chemicals can act as molecular machines to carry out the physiological reactions of life. For simplicity, I refer to these two programs as "genetics" and "structural biology." The so-called structural school played an important role in the genesis of molecular biology, but it was predominantly British in location, and its influence was strongest during the later years of this account when x-ray crystallography of macromolecules began to make major breakthroughs.[1] These two research traditions came together in the early 1950s when the genetic function and the physical structure of DNA (and a few other molecules) allowed a unifying explanation of both questions.

This book is about the first program, the one that led to an understanding of the material basis of heredity; it is about the adoption of bacterial viruses, or bacteriophage, to seek the physical nature of the gene. This program was developed by a group of scientists, primarily in the United States, that became known as "the American Phage Group" (APG). Starting in the late 1930s, several scientists independently started work on bacteriophages—only recently discovered in 1917—attracted by their apparently simple biology. Bacteriophages were characterized by their small size (smaller than bacteria, invisible under the light microscope), their presumably simple biology, and their ability to parasitize bacterial hosts and rapidly reproduce inside the bacterial cell, burst the host cell, and initiate another round of infectious reproduction. The ability of bacteriophages (phage, for short) to reproduce rapidly and faithfully made them ideal agents to use in the study of heredity. While "the nature of the gene" was a contested notion in the 1930s, a few scientists saw phages as essentially "pure genes."

Two concepts appear throughout this book, genes and reproduction, because they were the central theme of the scientific program of the APG. Both concepts, however, changed significantly, some would say revolutionarily, over the period involved in this story. The notion of "the gene" grew out of the nineteenth-century recognition of the regularities in inheritance and the need for some mechanistic explanation of these regularities—the "laws of inheritance." Until the middle decades of the twentieth century, the concept of the gene was unstable, with a quite wide spectrum of views on what it meant. For many, it referred to some (as yet unidentified) unitary particle of the cell that somehow controlled what the cell did, both during embryonic development and in the transmission of "traits" from parent to offspring. By the 1930s, some biologists had become committed to looking to physics and chemistry for new approaches to old problems. The geneticist Hermann J. Muller (1890–1967) spoke to a group of Soviet physicists in 1936 on the problem of the gene:

> The evidence obtained by geneticists indicates that it is in the tiny particles of heredity—the genes—that the chief secrets of living matter as distinguished from lifeless are contained, that is, an understanding of the properties of genes would bridge the main gap between inanimate and animate. Such a study would be of intense interest from the point of view of physics as well as of physical chemistry and organic chemistry for it is already known that these genes have properties which are most unique from the standpoint of physics and of the sciences related to physics. . . . It is in this field of mutation that the physicist is today most actively and fruitfully helping the geneticist. . . . The geneticist himself is helpless to analyze these properties [of the gene] further. Here the physicist as well as the chemist must step in. Who will volunteer to do so?[2]

For others, the gene was more nebulous, being, in the words of Richard Goldschmidt (1878–1958), "not an assembly of independent units but a kind of microorganism in which all parts are needed and interacting in some way, so that the entire chromosome, not definite parts, has to be called the genic material."[3] In an important historical and philosophical study of the concept of the gene, the philosopher Philip Kitcher examined the diverse ways that scientists refer, sometimes within one sentence, to quite different real-world objects as "genes."[4]

In one breath, a geneticist means "a sequence of nucleotides in a DNA polymer" and later in the same sentence means "the factor that explains a human being's blood disorder," sliding unconsciously between various technical meanings of "gene" with ease and complete understanding by her scientific audience.

As a result of phage experimentation combined with results from structural biology, by the early 1950s there was a rather abrupt, nearly universal acceptance of a detailed conception of the gene based on the idea that the linear chain of the DNA molecule was the physical gene, that the sequence of subunits (nucleotides) that were linked together to make the DNA chain formed a coded message or instruction that carried the information needed to explain parent-to-progeny heredity as well as that information needed to direct cellular functions in some orderly way. This explanation was so well confirmed, so elegant, and so powerful in an explanatory sense that from the mid-1950s it was the basis for almost all genetic research and theorizing.

The other concept, less dramatic in its evolution, was that of reproduction. The biologists, including phage workers, spoke of organisms "reproducing." Now, the concept is shared with the word "replicating," but at the time the APG was forming, its goal was framed in terms of phage reproduction. This distinction seems trivial, perhaps, but hidden in each word are assumptions about the suspected underlying mechanisms. Reproduction suggests a process similar to that involved in having children, breeding rabbits, and growing sweet corn. Replication, however, suggests mechanistic processes at a detailed micro level, such as running an offset printing press, making a bronze cast of a statue, or following a recipe in the kitchen. Early phage biologists talked about reproduction. Only later, as biochemistry, with its known mechanistic reactions, explored phage growth and reproduction, did it appear appropriate to speak of phage replication.

The Analytical Framework

The story of the APG—how it formed, how it developed as a research community, and how it proved to be a contributor to the formation of molecular biology—is the subject of this book. The concept of a "discipline" is a fraught one, with no universally accepted definition, yet it is useful as a way to characterize various coherent intellectual enterprises. This concept, however fuzzy, is used to organize educational, research, and social structures. A general taxonomy of disciplinarity would be complex: the philosopher Barbara Von Eckardt, for example, views "cognitive science" as a discipline in and of itself, yet a review of that discipline in 1978 asserted that cognitive science encompasses six other recognized disciplines: philosophy, psychology, linguistics, computer science, anthropology, and neuroscience.[5] Molecular biology, in the period considered in this book (1940–1960), was certainly an immature discipline, evolving over time. It might be argued that molecular biology started as a subdiscipline during this period and evolved to its hegemonic position in subsequent years so that now all biology is "molecular."

I believe that this latter view is an overstatement: field studies in ecology, for example, may use the techniques and methods of molecular biology, but their explanatory accounts are not couched in molecular terms. Molecular biology is based on molecular epistemologies, whereas there are great swaths of biology that are not. Often the term "interdisciplinarity" has been invoked to describe the origin of a new discipline, but that term begs for explanation. Disciplines that are already well established do not just come together, dissolve, and emerge as something new. Well-known examples include the emergence of psychology from fragments of philosophy and physiology, and of computer science from aspects of mathematics and electrical engineering.

This book provides a historical account of the individuals, possible mavericks, with diverse scientific backgrounds who came together, strategized, and managed to negotiate from starting ideas a set of common beliefs about phage research and then to attract and organize a group of followers who became the APG. This process, with both intellectual and social assumptions, led to the formation and evolution of the discipline of molecular biology as a natural inheritor of the basic shared assumptions of the APG.

My analytical framework, which goes beyond the vagueness of simple "interdisciplinarity," has been adopted and adapted from that of Von Eckardt's analysis in 1993 of the origin of cognitive science as a new discipline, a model she acknowledged as basically Kuhnian and which she calls the Research Framework Model. In her view, an immature science is one that lacks well-entrenched theoretical and methodological principles yet exhibits a shared commitment to a research framework nonetheless. A "research framework" in her analysis consists of four basic elements: first, a set of pre-theoretic assumptions that function to specify the domain of the scientific activity (*the domain element*); second, a set of pre-theoretic basic research questions, more or less known in advance of the research in the field (*the questions element*); third, a set of substantive assumptions about the approaches to be used, or what counts as answers to the research questions (*the answers element*); and fourth, a set of methodological assumptions as to how to carry out the agreed-upon approaches (*the methodology element*). "Research frameworks" in immature sciences are subject to revision and may exist in different versions at different times (*the amendment and revision strategies*). One task of the historian might be to identify these elements and their origins and to trace their evolution and revisions over time in light of new knowledge and intellectual challenges.

This book first examines in some detail various versions of these four elements of the "research framework" that seems to define the APG. After the examination of these individual components, the concluding chapter presents a more general synthesis of these versions. I provide signposts throughout the book to remind the reader how this analysis is being deployed and why particular historical evidence is relevant.

Existing Historiography

Most accounts of the history of molecular biology implicitly recognize two periods, roughly demarcated by the detailed understanding of the physical and chemical structure of the gene and its function in heredity, which was achieved in the 1950s. Before this divide, molecular biology was defining itself and its "systems"; after this period, the field has been devoted to exploiting and refining its basic paradigms.

The early history of molecular biology—pre–Watson and Crick and DNA, that is—has received little attention except for studies of specific landmarks such as the work of Robert Olby and Horace Freeland Judson. Recent accounts by Ulf Lagerqvist and Joseph Fruton are general descriptions with individual points of view. The canonical origin account of molecular biology as having been born of the mind of "The Physicist" persists in the lore of this discipline, and this account has been reinforced and legitimated by presentist accounts and persistent recollections of participants in celebratory volumes honoring the anointed founders. A recent, more balanced account from Michel Morange encompasses the European as well as the

American traditions. Biographical studies of several influential participants have contributed specific personal experiences and viewpoints to the history of molecular biology, beginning in 1968 with the (in)famous *The Double Helix* by James Watson (b. 1928).[6]

A remarkable collection of essays by the followers of Max Delbrück (1906–1981) was assembled in 1966, on the occasion of his sixtieth birthday, by three of his associates: John Cairns (1922–2018), Gunther Stent (1924–2008), and Watson. This volume of essays, *Phage and the Origins of Molecular Biology*, known colloquially as *PATOOMB*, has come to be regarded as the "official" history of the APG and by extension the history of molecular biology. These recollections have been taken as primary data, for the most part quite uncritically, by historians (such as Donald Fleming and Bernard Bailyn), sociologists (Nicholas Mullins), and scientists (Robin Holliday and Max Perutz). The canonical stature of this festschrift has been reinforced by the publication of a second edition, with a new introduction by John Cairns.[7]

The motivations behind *Phage and the Origins of Molecular Biology*, both public and private, both conscious and unconscious, are complex, but surely the volume is neither comprehensive nor analytical in its approach to the history of the APG in particular or molecular biology in general. At one level it is simply recollections by Delbrück's friends on the happy occasion of a milestone birthday. There was no requirement for documentation to support warm reminiscences. There was no need to dwell on the work of others. And certainly there was no need—indeed it would have been impolite—to mention any shortcomings of the principal celebrant. Yet this collection of reminiscences remains an important, interesting, and valuable contribution to the history of molecular biology, especially when taken in its unique context.

The Argument

Between the discovery of bacteriophages in 1917 by the French-Canadian scientist Félix d'Herelle (1873–1949) at the Pasteur Institute in Paris and the end of the 1930s, phages were objects of curiosity, medically noteworthy as potential anti-microbial agents in the era before antibiotics, but of fundamental interest to very few other scientists. The physical nature and biological properties of something that was invisible, killed bacteria, and apparently reproduced very rapidly inside bacteria did not elicit much attention among biologists of the time. Complexity was perhaps seen as more likely than simplicity to lead to biological understanding.

In the 1930s, however, several scientists of a more independent, contrarian, or maverick bent were pushing forward the study of phages, these apparent viruses of bacteria. Chapter 2 provides a brief survey of this work to set the stage for the more focused and groundbreaking work that was to follow. The scattered research of the 1930s offered the clues and the basis for the scientific ideas of the APG founders. This research provided evidence that phages are particulate objects, not "contagious fluids" as had been suggested earlier; that phages acted as exogenous agents to enter the bacterial cell and somehow multiplied inside the cell, which then ruptured and released newly made progeny phage; and that phages existed in types or strains with stable, identifiable hereditary properties. Beyond such very general facts about phages, much was unknown. However, phages were ubiquitous, easy to isolate from many sources wherever bacteria were found, easy to manipulate in the lab, rapid to grow, and inexpensive to study.

A second element that was to become crucial to research on phage by the APG, influential in the first half of the twentieth century but now nearly forgotten in biological research, is described in chapter 3: the importance of radiation biology in the study of cells before the biochemical revolution of post-war science. Derived from the atomic scattering experiments of J. J. Thomson (1856–1940), the use of radiation to bombard cells and organisms to perturb biological functions provided a widely applicable experimental probe of biological processes at a fundamental physical level, based on the interaction of the radiations with the chemical components of the biological system being studied. A theory of radiation as "projectile" and the cell components as "targets" was developed in the 1920s and 1930s to analyze radiation effects on living things. This so-called target theory became a mainstay of avant-garde biological research, especially for work with poorly understood concepts such as the gene, protein synthesis, and reproduction. It was, initially, the main experimental tool that united much of the nascent APG community and was an essential element of its research framework.

In addition to the background knowledge of phages and the development of the key methodological tool provided by radiation biology, it is important to survey the diversity of individual scientists who contributed, centrally as well as peripherally, to the nascent APG. Chapters 4 and 5 present a collective biographical study of some of the individuals and scientific traditions of these early participants. Since this book is not a simple narrative of the so-called founders of the phage community and their acolytes, these two chapters emphasize the diverse intellectual commitments, personal stories, and scientific motivations of a rather broad and heterogeneous collection of scientists, brought together by their common interest in phages as an object of research.

Having given the contexts that existed at the beginning of the 1940s, the first five chapters set the stage for the initial formation of the APG, the coming together of a small founder group, which I term the nucleation, in chapter 6. Here we first meet the key organizers of the phage group: Max Delbrück, Salvador Luria (1912–1991), and Alfred Hershey (1908–1997). They were by no means the only phage workers at the time, but they do deserve a higher place in the phage pantheon because of their crucial role and sustained focus in initiating, maintaining, and advancing the organization of the community that became the APG. They were the ones who influenced the research framework that was the core of the APG and that ultimately shaped the contours of the new discipline of molecular biology. This chapter shows how the four elements of the research framework that came to characterize the APG were codified. In succinct form they can be stated as follows:

1. The *domain element:* reproduction, especially the nature of the gene and its remarkable stability.
2. The *questions element:* what is a gene and how can it be reproduced with high fidelity, but be equally faithfully reproduced when a rare mutation happens?
3. The *answers element:* the answers will be given in terms of physical principles based on chemistry and physics of cell components.
4. The *methodology element:* simple experiments, with observable and quantitative outcomes, amenable to theoretical models and analysis are to be preferred. Mathematical analysis and quantitative predictions are encouraged.

We will follow these four aspects, in various levels of detail, through subsequent chapters to see how they formed the organizing framework for group identity and research programs within the APG. The related but different issue of how the research framework responded and adapted to challenges (*amendment and revision strategies*) is considered in chapters 9 and 10, devoted to two major problems in phage research: the phenomenon of lysogeny, in which the simple model of exogenous phage infection is complicated by the existence of a latent, resident intracellular form of the phage in some bacteria, and, a bit later, the discovery of a diversity of phage types, with differing forms of genes (RNA versus DNA, circular versus linear genomes), which challenged the simplicity desideratum that was initially sought. How a small formative group deals with challenges to its core principles is particularly important for the ultimate survival and success of its research framework. These challenges come in many forms—personal conflicts, intellectual disagreements, institutional pressures, and experimental tastes, to list just a few. The basic core commitments of the APG faced two significant challenges, as noted above: what to do about lysogeny and what to do about the discovery of new kinds of phages. The former was a challenge to the simplicity of the external phage infection model, and the latter was a challenge to the restriction of phage material to an agreed upon set of phages for the sake of inter-laboratory comparability, again a desideratum of simplicity. How these challenges were met, how the research framework evolved to accommodate the new findings, and how they enriched phage research and provided models for future evolution of the research framework are the subjects of these chapters.

With the initial nucleation of a few like-minded phage enthusiasts, strategies and efforts to recruit colleagues with similar commitments, interests, and goals became essential in order to establish a viable research community. In chapter 7, these community-building activities will be examined and shown to reinforce, propagate, and police the group adherence to the framework of shared commitments that characterized the growing APG. These activities included periodic small gatherings, the "Phage Meetings," at which group members could present their findings for critique and comment, receiving support from the group. Soon the Cold Spring Harbor Laboratory's summer "phage course" would be institutionalized to teach basic phage research techniques, present lectures on current phage research, and socialize new participants into the group norms. In effect, these activities were recruiting, indoctrination, and certification processes to build cohesiveness around the group framework of shared commitments. A further organizational device was an irregularly distributed newsletter, the *Phage Information Service,* that fostered communication and collaboration between the far-flung community that grew out of the phage course and the phage meetings, all centered around the Cold Spring Harbor Laboratory, which became the intellectual home for the APG. The particular nature of this venue and its role in the nurturing of phage research and subsequently molecular biology is described in chapter 8.

Although this book focuses on the phage research in North America, mainly the APG, this particular scientific community had close and strong relationships to several phage groups in Europe, especially at the Pasteur Institute in Paris and at the University of Geneva. How these centers of phage research interacted, both intellectually and personally, is the subject of chapter 11. The diffusion of the framework of shared commitments among these centers and the transnational aspects of phage research probably bled over into the nascent discipline of molecular biology as it developed on both sides of the Atlantic.

In addition to the elements of the research framework

that characterized and guided laboratory experimentation and intellectual output, there are social elements that characterize any group activity, and the APG was no exception. The extent to which social norms affect experimental and intellectual activities is a subject of immense current interest but also immense uncertainty. It is tempting to hypothesize that a group such as the APG operated with something akin to its research framework of shared commitments in the social realm as well—a kind of "framework of shared social commitments" that guided and governed behaviors such as credit sharing, authority, secrecy and openness, gender and racial equity, and interpersonal respect. Although it is still uncommon to incorporate such elements alongside scientific and intellectual historical analysis, it is illuminating to do so in the case of the APG in part because its social structure exhibited a distinct departure, even rebellion, from traditional biology at the time. Moreover, examining factors like gender and race—even in a preliminary way given the paucity of concrete data—foregrounds the need for future considerations of such issues. Chapter 12 provides a preliminary survey of what we know of these social factors at work in the APG.

As the APG's solutions to some of its key original questions about the nature of the gene, reproduction, and protein synthesis became mainstream knowledge, many of the early phage workers saw opportunities to apply their phage knowledge and research approaches to other problems in biology. This natural maturation took its toll on what had been a strongly cohesive and committed group of phage biologists. How this maturation, migration, and morphing of phage biology into molecular biology evolved is described in chapter 13. The reconfiguring and recasting of the APG's basic framework of shared commitments into those of molecular biology completes the account of discipline formation described in this book. Of a more general nature, chapter 14 recapitulates the utility of Von Eckardt's research framework model for discipline formation and its applications to the case of molecular biology.

2

Phage Before the War
(1917–1940)

"All I know is that it's hell on the bugs . . . dissolves them, eats them up, slaughters them, wipes them off the. . ." These were the words of young Dr. Martin Arrowsmith (played by the 1930s heartthrob Ronald Colman) describing his discovery of bacteriophage in *Arrowsmith*, a 1931 film that was nominated for an Academy Award. The movie was based on the Pulitzer Prize–winning novel of the same name by Sinclair Lewis (1885–1951) and described a young scientist's excitement at the discovery of the potent antibacterial agent called bacteriophage. The novel was based on the actual discovery of bacteriophage (phage) by Félix d'Herelle, a French-Canadian scientist working at the Pasteur Institute, who published his first report on phage in September 1917.[1] As in the Hollywood film, the real-life discovery set off a wave of interest and exploration of phages as useful anti-infectious agents in an age before antibiotics when there were precious few effective treatments for common infections.

In research on phages, from their discovery in 1917 to the beginnings of the APG in 1940, a few intrepid and scattered scientists tried to understand phages as biological objects; more often, however, scientific investigation was directed toward the therapeutic potential of this new anti-infectious disease agent. This chapter reviews and evaluates the meager basic knowledge about phages that was accumulating during these decades; this background gives us the starting point for our analysis of the APG and its role in the development of the discipline of molecular biology.

From the discovery of bacteriophage in 1917 until the disruptions and dislocations of World War II we can trace the trajectory of phage research through an analysis of the scientific literature that specifically dealt with this field. With admirable thoroughness, Hansjürgen Raettig (1911–1997), a professor at the Robert Koch Institute in Berlin, cataloged the world literature on phage from 1917 until 1956.[2] With his data we can capture, in a quantitative way, the focus on phage and the topics in phage research that attracted the attention of scientists in the interwar period.

From Raettig's graphs and numerical analyses, we find that from 1917 to 1940 (inclusive) there were some 2,480 publications on phage. These include original contributions, reviews, translations, and summaries, so this number certainly overstates the quantity of original research in the field. However, the breakdown into subfields as a fraction of the total is probably more informative of the attention being paid to specific aspects of phage biology. Raettig used nine categories to classify these works in the phage literature: foundational studies, phage epidemiology, phage therapy, immunity, morphology, lysogeny, reproduction, biochemical studies, and genetic studies (table 1). As might be expected, work on the nature of bacteriophage and phage therapy were main topics of interest, yielding together almost half (47.3 percent) of the published literature on phage during this period. Since the electron micro-

Table 1. Analysis of Phage Literature
Through 1940 by Category

Category	Number of Citations	Percentage of Total
Foundational	616	24.8
Epidemiology	176	7.1
Therapy	557	22.5
Immunity	163	6.6
Morphology	42	1.7
Lysogeny	116	4.7
Reproduction	196	7.9
Biochemistry	299	12.1
Genetics	315	12.7
Total	2,480	100

Source: From Hansjürgen Raettig, *Bakteriophagie, 1917 bis 1956*, 2 vols. (Stuttgart: G. Fischer, 1958), 20–24

scope was invented at the end of this period and was only beginning to be applied to phage in 1939, morphological studies were rare (1.7 percent). Interestingly, the topic that was to become probably the most fruitful area in the subsequent decade, phage reproduction, represented only a small part (7.9 percent) of the phage literature in this period.

Although the bacteriophage phenomenon interested many microbiologists and was treated extensively in the major textbooks of the period, there were relatively few places where serious phage research was being done.[3] Most of this work was aimed at the potential of phage therapy, using phages to treat infectious diseases. Because the presence of phages seemed to increase in patients just as they were recovering from their diseases, d'Herelle speculated that phages were responsible for the recovery from infectious diseases (he called it exogenous immunity), and hence phages would be useful, if isolated, as therapeutic agents in such cases. In Europe, phage research

centered around the Pasteur Institute, its historical origin. Although d'Herelle left Paris in the early 1920s for Leiden—later
Alexandria, Egypt, and eventually America—two of his colleagues, Eugène Wollman (1883–1943) and Elisabeth Wollman
(1888–1943), developed a strong research program in phage
biology at the Pasteur Institute, a tradition that persisted even
after they were victims of Nazi genocide. Another hub of European phage research was the National Institute for Medical
Research in England, located at Mount Vernon Hospital (relocated in 1950 to Mill Hill), in the group headed by the British virologist Christopher Andrewes (1896–1988). It was in his
laboratory that some of the first studies on the biophysics of
phages were conducted and that the Australian and future Nobelist Frank Macfarlane Burnet (1899–1985) got his start on
phage work in the 1930s. This was also where the first biochemical studies of phage were initiated by the young Hungarian-
German refugee Max Schlesinger (1907?-1937). When Burnet
returned to Australia, he established an important center of
phage research that had major influences on later phage work
as well as virology in general.

In America, phage research in the interwar period centered on three institutions, with other scattered work beyond
these centers: the Rockefeller Institute, Stanford University,
and Yale University. Due to the transplantation of individuals
from these centers, phage research sprang up at Washington
University (St. Louis) at the very end of this early period. In
other cases, here and there, a curious scientist would begin to
study phage in his or her own way, based on reading the literature and employing the simple microbiological methods that
were all one needed to get started. This was precisely how
phage came to the California Institute of Technology (Caltech),
which was destined to become the center of the American Phage
Group.[4]

It was probably d'Herelle's work on phage therapy that first attracted most bacteriologists' attention to phage. Before the antibiotic era, any potential therapy for infectious diseases drew notice, and the simplicity of the techniques of phage research meant that it could be taken up by any competent bacteriologist. Indeed, phage therapy seemed to be a popular subject for beginning researchers and medical student theses.[5] Phage research at the Rockefeller Institute was probably initiated by André Gratia (1893–1950), a young post-doctoral visitor from Brussels and a protégé of the Nobelist Jules Bordet (1870–1961), director of the Pasteur Institute in Brussels. Bordet was one of the earliest phage researchers in Europe, partly due to his Pastorian connection and partly because of d'Herelle's challenge to Bordet's conception of immunity. Bordet was unconvinced by d'Herelle's notion that phage played a major role in recovery from infectious diseases, a recovery that represented a sort of "exogenous" basis for immunity and a direct challenge to Bordet's serological immunology. Gratia had come to work with Simon Flexner (1863–1946), and he defended Bordet's position in the controversy, starting in 1921, between d'Herelle and Bordet. At this same time (1923) the Ukrainian immunologist Jacques Jacob Bronfenbrenner (1883–1953) joined the Rockefeller Institute as an associate member and initiated a research program on bacteriophage. Bronfenbrenner and his colleagues Donald M. Hetler (1896–1956), G. M. Kalmanson (1915–1994), and Charles Korb (b. circa 1900) published thirty-eight papers (as listed by Raettig) on the physical properties of phage, sensitivities to various inactivating treatments, the volatility of phage, and phage therapy in animal models. Some of these early papers contain references to microscopic filming of real-time bacterial lysis by phage. Unfortunately, these films seem to have been lost. Bronfenbrenner's work on phage would appear to have been a continuation of his earlier interest in antiviral therapies in research carried out as a student in Paris with Alexandre Besredka (1870–1940) at the Pasteur Institute between 1907 and 1909.[6]

In 1928 Bronfenbrenner, along with another virologist from the Rockefeller Institute, Edmund V. Cowdry (1888–1975), joined the department of bacteriology and immunology at Washington University in St. Louis, with Bronfenbrenner as chair. Bronfenbrenner's wide interests extended to matters of public health as well as basic research, and he recruited scientists of real talent to his department, one of whom, Alfred D. Hershey, would be one of the key founders of the APG.

A closer examination of phage research in the interwar period reveals a conceptual divergence between the medical and the biological aspects of bacteriophage. The promise of phage as antibacterial agents of therapy was certainly the more prominent of these themes. Interest in the biological nature of phage, while attracting less attention, proved to be both more contentious and more fruitful of new insights. It is this thread of interwar research to which we will now turn to find the roots of the APG.

Although Félix d'Herelle fervently hoped to follow the Pasteurian path to discoveries of great medical import, he found himself distracted by serious challenges to his concept of bacteriophage as an "ultravirus" or an obligate intracellular microbe. For about ten years, he was engaged in a very public debate with Jules Bordet and André Gratia on the nature of bacteriophage. This debate has been the subject of several scholarly studies that have identified key reasons for these differing views: the historian Ton van Helvoort has argued that there was a basic difference in scientific "style," with d'Herelle favoring a "bacteriological style" and Bordet and his school preferring a "physiological style."[7] In addition, d'Herelle's status as an outsider coupled with his often "irascible" personal attacks on his opponents seemed to help make him the target of attacks by Bordet and Gratia. The authority of Bordet, as director of the Pasteur Institute in Brussels and as a recent Nobel Prize winner, initially attracted leading microbiologists to his view that "the bacteriophage phenomenon," as it was called, represented an activation (or "vitiation") of a cellular function that produced a lytic substance capable of serial activations of other cells. The major textbooks of the interwar period were generally supportive of the Bordet view or carefully agnostic. None could be said to be clearly on d'Herelle's side.

Although there were many articles on the nature of phage at this time (in Raettig's classification there were 616 publications, or 25 percent, on "foundations"), they tended to be polemical rather than experimental. More of these were published between 1920 and 1930 than in the following decade, suggesting both a waning interest and a lack of resolution of the disputes. Of course, with the visualization of phage in the electron microscope in 1939, the controversy over the particulate nature of phage was effectively settled, with d'Herelle's concept being vindicated. But it was precisely these debates on the nature of bacteriophage that would attract the serious concerns of scientists who saw in phage a glimmer of hope to unlock deep biological problems of life and reproduction. It is this small group of scientists who would direct the field in new directions, including the formation of the APG.

In Paris, after d'Herelle's departure from the Pasteur Institute in 1923, phage research was sustained by the young husband-and-wife team Eugène and Elisabeth Wollman. The Wollmans published extensively between 1921 and 1940 (Raettig lists eighty-eight citations) before they fell victim to the Nazis. Their work focused almost exclusively on understanding the basic biological properties of phage, their interactions with host bacteria, and the confusing phenomenon of lysog-

eny. The Wollmans wholeheartedly adopted d'Herelle's view
of phage as a filterable virus, and apparently they had cordial
relations with d'Herelle's family, who lived in the same apartment building in the 1920s.[8] In the 1930s, radiation became an
essential tool for the Wollmans, as it did for others who studied the very small in both atoms and cells. Almost next door to
the Pasteur Institute was the world-famous Radium Institute,
where the biological effects of radiation were of keen interest.
Marie Skłowdowska Curie (1867–1934) herself had undertaken
radiobiological studies of bacteria in the 1920s, and the poly-
math French physicist and resistance hero Fernand Holweck
(1890–1941) was dabbling in radiation biology among his many
interests. Holweck and his young Italian associate Salvador
Luria collaborated with Eugène Wollman to use x-rays to estimate the size of a bacteriophage in 1940, initiating a phage
research pathway that would branch in several fruitful directions.[9] Luria would play a key role as one of the founders of the
APG, so it is important to trace his history in this period, too.

Salvador Luria was born in Turin into a lower-middle-
class family with a famous surname dating back to the beginning of Lurianic Kabbalah in the sixteenth century. He was
educated as a physician, but he had little enthusiasm for medicine in the tumultuous 1930s in Italy. In one of those contingencies recognized as important only in retrospect, Luria's
boyhood friend Ugo Fano (1912–2001), who studied physics,
convinced him of the exciting things going on in that field; as
a result of Fano's machinations, Luria went to Rome to become
a radiologist and to study physics at the same time. It was this
"year among the physicists" that introduced Luria to radiation
biology and, especially, the work by Max Delbrück on the nature of the gene as molecule.

A further coincidence made him aware of the obscure
microbes called bacteriophage: on a stalled Rome trolley in

1938, Luria struck up a conversation with a fellow passenger who turned out to be Geo Rita (1911–1994), a biologist studying the phages in the Tiber River. Luria later attributed this conversation as his first introduction to bacteriophage. When the young man decided to leave fascist Italy in 1939, his thoughts turned to the few places that might appreciate his odd combination of education and interests.

The Radium Institute under Holweck turned out to be just the place, and in 1939 Luria went to Paris. The fact that he had a strong letter of support from Enrico Fermi (1901–1954) in Rome probably did not hurt. Although Luria remained in Paris less than two years, leaving just ahead of the Nazi invasion in 1940, he consolidated his interest in phage research and honed skills and approaches that would help define his later phage work and the formation of the APG. Both radiation biology and statistical reasoning associated with the target theories, which we will examine later, were key tools that characterized phage research in the following decades.

With Luria's departure from France, the deaths of the Wollmans and Holweck, and the disruption of the French scientific establishment by the war, phage research in Paris came to a virtual halt. Only after the war would it be revived—significantly, the result of collaborations between Elie Wollman (1917–2008), the son of Eugène and Elisabeth, and the phage group at Caltech, as well as a postwar fellowship that allowed the young French scientist Raymond Latarjet (1911–1998) to visit the United States and, by dint of their Radium Institute acquaintance and by fortunate circumstances, to collaborate with Salvador Luria.[10]

While radiation biology was the main approach to understanding the nature of phage in the interwar period, some research that could be called biochemical was in its initial stages in both Paris and Germany. The well-known colloid chemist

Heinrich Bechhold (1866–1937) and his younger Hungarian colleague Max Schlesinger in Frankfurt am Main were among the first to apply chemical analyses to phage. To do this, they pioneered ultracentrifugation methods for purifying phage from crude lysates. In a series of publications between 1930 and 1937, Schlesinger described the sizes of the phages he studied (which included S13, a phage ultimately used by the APG), as determined by the new sedimentation methods of the Swedish physical chemist Theodor Svedberg (1884–1971), as well as the chemical reactions of purified phage. Two striking results were the high phosphorus content of phage (3.7 percent) and the high affinity for basic dyes, which Schlesinger interpreted as indicating "the nucleoprotid" nature of phage. In 1935 Schlesinger fled to Great Britain and joined William J. Elford (1900–1952) and Christopher Andrewes at the National Institute for Medical Research in London, but he was reputed to suffer from depression, and, sadly, he ended his life and promising start in phage research at age 30.[11] In September 1936, only a few months prior to his self-inflicted death, Schlesinger published his most definitive evidence that phages contained DNA because of their strong Feulgen reaction, a chemical test believed to be diagnostic for DNA (called thymonucleic acid at the time because the usual source was nuclei of thymus cells). He noted as well that "the phage-substance seems to be chemically different from any constituent of the bacterial cell normally present in significant amounts."

In Paris, the interest in chemical and physical study of phage was subordinate to the biological interests of the Wollmans, but in the early 1940s Paul Bonét-Maury (1900–1972), Vladimir Sertić (1901–1983), and Nicolai Boulgakov (1898–1966) focused on another small phage, φX174, for their measurement of the size of phage by both radiation target analyses and sedimentation studies.[12]

Interwar phage research in Britain was concentrated in the laboratory led by Christopher Andrewes in London, where Andrewes and Elford published a handful of papers that formed the basis for the physical characterization of phage by sedimentation and filtration methods. They were joined by their first research fellow in 1925, the young Australian Macfarland Burnet, who would soon develop his own phage research program in Melbourne. While with Andrewes and Elford in London, Burnet developed and refined his ideas about the nature of phage, and by the time he returned to the Walter and Eliza Hall Institute in Australia he had a clear agenda for his phage work. Although he was medically educated as a pathologist, he found the precision of phage research much to his taste, and his publications through the 1930s did much to clarify the phenomenon of lysogeny, the puzzling case where it seemed that phages appear from nowhere in certain bacterial cultures. With his colleague Dora Lush (1910–1943), Burnet showed that the mystery of lysogeny appeared to be a property of the phage, not the bacterial host, because they found related phages that seemed to differ only by a mutation that controlled the ability of the phage to produce lysogeny. With this important understanding, a major stumbling block in phage biology was on its way to removal. This work on lysogeny also established Burnet as one of the key phage researchers of the interwar period.[13]

While the majority of phage research in the interwar period was medical in focus because of the promise of phage therapy for bacterial diseases, a few scientists were exploring the fundamental biological nature of phages. The experimental tools to study viruses as biological objects were frustratingly few. Prior to the development of electron microscopy in the late 1930s, the mainstay approach of the biologist, morphology, was not available to virologists. Chemistry, a relatively new tool for biologists, was useless until pure, well-defined samples of viruses could be obtained. Viruses did not seem to have observable metabolism to study, and they did not catalyze chemical processes such as fermentation, as did more complex microbes such as yeasts and bacteria. Virologists were limited to observations of effects on the host organisms the viruses preyed on, such as disease, and the fact that host animals often produced specific antibodies to viruses that could be observed in a few inactivation assays.

What virology looked like to the interwar biologist was a lab bench with an assay system; the main question one could ask was "is it active or is it not" in a particular assay. This simple binary outcome, after any particular manipulation, was, for many, the main reason for avoiding virology. However, for a select few physicists who were used to inferring complex, unobservable phenomena and entities from just this sort of "yes/no" output, viruses seemed to provide a clarifying simplicity. This was the kind of work many of them had been doing to elucidate the complexity of the atom in the early days of the twentieth century. Biology looked ripe for their insights, models, mathematics, and clarity of thought. The next chapter considers how some of these themes came together to ignite mid-century phage research.

3
From Physics to Biology
The Target Theory

Genes are mysterious things. From their earliest days in the nineteenth century, their very existence as "things" was debated. Did they have a physical existence or were they some sort of conceptual entity without physical reality? How could they have such stability yet abruptly change in one generation in the life of a living organism? Even scientists who might be called "geneticists" could doubt their reality.[1] Even though genes, whatever they are, seemed to associate with physically real, visible cell structures, the chromosomes, it was still unclear just what the gene was and how it exerted its effect to determine cell growth, differentiation, and function. As a fundamental biological problem, genes attracted very little medical interest in the first half of the twentieth century. It was not until the nature and function of genes, per se, were clarified that medical genetics became respectable. So genetics was left to the theoretically inclined folks and the practical agricultural breeders. The former includes physicists and mathematicians who would, as it turned out, contribute substantially to the origins of molecular biology.

Radiation effects attracted other biologists in addition to geneticists: some experiments were interpreted as showing a stimulatory effect on cell division by radiation emanating from adjacent living cells, so-called mitogenic radiation.[2] In the work of John Butler Burke (1873–1946?), the mysterious objects (radiobes), resembling cells, that formed in gelatin containing uranium salts were taken as evidence that the radiation emanating from the uranium was involved in the origin of life.[3] Some were more fortuitous: in his laborious search for ways to increase the yield of mutants in his fruit fly stocks (other than just waiting for one to appear), H. J. Muller found in 1926 that x-ray treatment of male flies prior to mating dramatically increased the frequency of mutations affecting diverse genes, a discovery that led to the Nobel Prize in Physiology or Medicine in 1946.[4] Muller's finding would lead directly to the widespread application of ionizing radiation to basic genetic research.

In the canonical account of the origin of molecular biology, a paper with the promising title "On the Nature of Gene Mutations and Gene Structure" (English translation) was presented to the Göttingen Society of Science, and thus was born molecular biology. This paper, presented in April 1935, outlined Max Delbrück's theory of the gene, a theory about which Erwin Schrödinger (1887–1961) pronounced in 1944: "If the Delbrück picture should fail, we would have to give up further attempts."[5]

What was so important about this paper and why has it become a "landmark"? Discounting the post hoc canonization of early work by the historical "winners," there are two more significant aspects of this paper that have caught the imagination of scientist and historian alike. First, it is a bold new attempt to tackle a biological problem with a new set of tools, those of physics. Second, the paper represents a conscious collaboration of a geneticist, a biophysicist, and an atomic physicist. Such a collaboration was seen by the three participants as an opportunity to bring different disciplinary forces to bear on a problem of common interest. One of the authors, the bio-physicist Karl Zimmer (1911–1988), later recalled: "its friends and critics used to refer to it as the 'Green Pamphlet' [because of the bright green cover on the offprints] or, somewhat more deprecatingly, as the 'Driemännerwerk' ('Three-men-paper'): team work was not very usual in Germany thirty years ago, and inter-disciplinary team work appeared rather strange to some scientists."[6] Nikolay V. Timoféef-Ressovsky (1900–1981), the geneticist, outlined the biological problem: what is the nature of the mutation process and what is the nature of the gene? Zimmer outlined an approach to the structure of the gene using physical experiments based on the recently discovered action of x-rays in causing mutations. This approach was called the "target theory." Delbrück used the evidence and concepts from the target theory experiments to construct a model of muta-tion and then a theory of gene mutation and structure—in Gunther Stent's words, "a 'quantum mechanical' model of the gene."[7]

What is the target theory, how did it originate, and what is its current status? The "target theory" is the name given to a model for the way radiation interacts with cells. The two basic features of this model are (1) radiation is considered to behave as random projectiles, and (2) the components of the cell are considered as the targets to be bombarded (and inactivated or otherwise modified in some observable way) by these projec-tiles. From the nature of the dose-response relationship for a specific type of radiation, and inactivation of a specific biolog-ical function, the number and size of the subcellular targets for that particular function can be calculated. This approach has been used to estimate the size and shape of enzymes, viruses, and, as in the Three-men-paper, genes. It has also been applied to study the subcellular apparatus that synthesizes proteins and DNA, as well as such global physiologic processes as respiration and ion transport. From the mid-1930s to the mid-1950s, the target theory and its applications were a major preoccupation of the field that became known as biophysics.[8] It is hard to overemphasize the role that radiation biology and various versions of the target theory played in the origins of molecular biology and of phage research in particular.

The earliest and most frequently cited paper that describes the target theory was written by James Arnold Crowther (1883–1950) and published in 1924 in the *Proceedings of the Royal Society.* Crowther is universally credited as the originator of the target theory as it is known to biologists. Nonetheless, the fundamental notions involved in the target theory had been employed by physicists for over a decade. To better understand how the concepts from one field are appropriated and used in another field, it is useful to examine in more detail the circumstances of the origin of the target theory in biology. Various words and metaphors have been used to try to capture the essence of this process of scientific concept movement: "diffusion," "migration," "appropriation," "transfer," and the like. All these words and metaphors seem to come loaded with more meanings and corollaries than are warranted. Is it an active or passive process? A push or a pull, and so on? If we consider the events in this case, taken from the historical record, we may be able to infer something of the conditions and processes involved in the transfer of basic concepts of atomic physics to applications in biology.

Atom Smashing

A brief summary of the relevant physical studies will set the context of this investigation. Around the turn of the century

Joseph John (J. J.) Thomson and his colleagues in England and Henri Becquerel (1852–1908) and his colleagues in France were interested in the nature of the newly discovered radiations emanating from uranium. One outcome of their investigations was the discovery that there were two kinds of radiation emanating from uranium, distinguished on the basis of their penetration of matter: α (alpha) rays were absorbed by very thin layers of material, while β (beta) radiation was more penetrating. These discoveries led, in turn, to an interest in understanding the absorption process itself—that is, the ways that the radiation interacted with the absorbing material.

In 1895, Wilhelm Conrad Röntgen (1845–1923) discovered a third type of radiation emanating from modified cathode ray tubes, which he called "x rays." Not only were these radiations absorbed but, it was found, they were scattered. In his useful and detailed description of the work of Thomson and Ernest Rutherford (1871–1937) on scattering of alpha and beta radiation, the science historian John Heilbron says that Thomson, a professor of physics at Cambridge, published in 1906 "one of the most important papers on atomic structure ever written."[9] This paper described Thomson's quantitative attempts to determine the order of magnitude of the number (n) of electrons contained in an atom of atomic weight A. Thomson's approach to the study of the internal structure of the atom was to employ the scattering of x-rays, the absorption of β-rays, and dispersion of light by gases. His object was to test a model advanced by William H. Bragg (1862–1942) and Rutherford that the atom consisted of a large swarm of moving electrons (the famous beehive model). This paper appeared the same month that Crowther received his bachelor's degree at Cambridge, joined Thomson's department, and took up the challenge of determination of (n). Between 1908 and 1912 Crowther published his results of a series of scattering experiments, which

significantly influenced Bragg and Rutherford in their thinking about the interior structure of the atom. In 1912 Crowther
became a demonstrator and lecturer in physics in Thomson's
department at the Cavendish Laboratory, a position he held
until 1924. In the context of our investigation of the target theory, then, Crowther's problem was to study the unseen, internal structure of the atom, and his approach initially focused
on the observations of the projectile—that is, the incident and
scattered particles and radiations, with apparently little attention to the effects on the absorbing or scattering material.

Crowther seemed to get on famously with Thomson and
was apparently an adept experimentalist. Then came the Great
War, and Crowther was detailed to serve in the medical radiography unit at Addenbrooke's Hospital in Cambridge, where
many war casualties were taken. Such facilities for x-ray diagnosis were rare, and Crowther took this opportunity to collaborate with medical colleagues. These contacts had a lasting
impact on him, and after the war he became involved in the
medical aspect of radiation studies. In 1921 he was appointed
lecturer in medical radiology at Cambridge, concurrently with
his position in the physics department. He wrote a text for the
Cambridge diploma course in medical radiology, *Principles of
Radiography*, and edited another book on applied topics: *The
Handbook of Industrial Radiology*. Although this switch in direction may have been stimulated by financial and professional incentives (according to Heilbron, Crowther "had long
wanted an independent, and needed a more remunerative position"), Crowther maintained these interests even after he was
appointed professor and chair of the physics department at the
University of Reading in 1925.

As already noted, the physicists studying atomic structure
initially focused on the fate of the projectiles—that is, absorption and scattering—but by about 1914 Rutherford and others

started to talk about experiments that might show the transmutation of elements by absorption of subatomic particles. A next logical step in the bombardment experiments was to look at the fate of the target material. Atomic theory suggested that it might be possible to transmute one element into another by bombardment with the right particles with the right energy. As early as 1910 there was talk of needing high-voltage technology to accelerate particles for bombardment experiments to overcome the limitation on the energies of the natural sources of particles, the α- and β-rays from uranium. The physicists therefore began to focus on the target and its fate under bombardment. By 1919 Rutherford had clearly formulated these notions in terms of target atom destruction—"atom smashing." In 1922, Rutherford's students James Chadwick (1891–1974) and Etienne Biéler (1895–1925) described a model of the nucleus as an oblate spheroid calculated from collision theory and based on particle bombardment and scattering.[10]

Crowther was, of course, immersed in this work. In successive editions of his popular textbook *Ions, Electrons, and Ionizing Radiations*, he described the bombardment and transmutation work with increasing detail and enthusiasm as the text went through rapid revisions from the first edition in 1919 to the fourth edition in 1924. For the fourth edition he added a new section titled "Collision of α-particles with atoms" with a long discussion of different target and projectile models.

During these same two decades three other results were widely heralded in physics, each of which contributed to the notion of projectiles and targets. In 1912 Charles T. R. Wilson (1869–1959) published his remarkable photographs of visible tracks of "ionizing particles" observed in his newly invented cloud chamber.[11] These photographs made visible the idea not only that particles collided with gaseous water molecules to serve as nucleation sites for droplet condensation, but that the particles could be "seen" to hit and ricochet from atoms and possibly other subatomic structures.

In this same year, James Franck (1882–1964) and Gustav Hertz (1887–1975) published the results of an experiment that was designed to test the ionization of molecules in the vapor phase by accelerated electrons. Their results were surprising, even to them. This was the first new test of the Bohr model of the atom with bound electrons. The so-called Franck-Hertz experiment also required a "projectile-target" mental construction.

A third important result in atomic physics of this period was the careful measurement of the wavelength of scattered x-rays by Arthur H. Compton (1892–1962). In 1923 Compton published his experiments, which showed that at high energies, x-rays were scattered not only by elastic collisions but by inelastic collision. While elastic scattering could be explained by a wave model for x-rays, the inelastic scattering was incompatible with the simple wave theory—for example, scattered light should not change its wavelength. Compton scattering was further evidence for the particulate nature of x-rays, and the model was clearly formulated in terms of "targets and projectiles." Incidentally, this work vindicated Einstein's much maligned view of the particulate behavior of light energy in general, since x-rays had been conceived only as very energetic light rays.

Radiation and Biology

In the early 1920s growing cells and tissues outside the animal body was a major goal of biologists studying cancer. One of the pioneers in this effort was Dr. Thomas S. P. Strangeways (1866–1926), director of the Cambridge Research Hospital. Since cancer was treated by exposure to x-rays and radium, Strangeways and others wanted to study the effects of radiation on cancer tissue and cells in isolation growing in culture flasks.

In 1923 Strangeways and a young radiologist, H. E. H. Oakley, published a paper titled "The Immediate Changes Observed in Tissue Cells After Exposure to Soft X-Rays," which caught the attention of Crowther. This paper gave qualitative observations on the frequency of tissue culture cells entering into mitosis as a function of duration of exposure to x-rays. The highly qualitative nature of these data is such that it is very hard to understand why anyone would be tempted to base any sort of quantitative physical theory on them. However, two factors may be relevant: first, the experiments used radiation bombardment, one of Crowther's primary interests; and second, because Crowther and Strangeways were both longtime fellows of St. John's College in Cambridge, they undoubtedly had opportunities to discuss Strangeways's work beyond the vague and qualified descriptions in his publications. Crowther wrote: "The authors find that after an exposure to the rays of 5 minutes the number of cells in mitosis was appreciably diminished, after an exposure of 10 minutes the number was still smaller, after 15 minutes only a few cells in mitosis were visible, and for still more prolonged exposures cells in mitosis were only seen occasionally in some of the cultures." He went on to state: "Although no numerical data are given these statements suggest *very strongly* that the number of cells capable of passing into mitosis is decreasing exponentially with the time of exposure; in other words that the action of the x rays on the cells which produces the incapacity for entering into mitosis follows a probability law" [emphasis added].[12]

For Crowther, "it seemed interesting to consider whether this probability might not be due to the x rays themselves, and represent the probability that a given structure in the cell would actually be affected by the incident radiation." From this point, Crowther went on to derive a simple expression relating the size of this "given structure" to its sensitivity to inactivation by

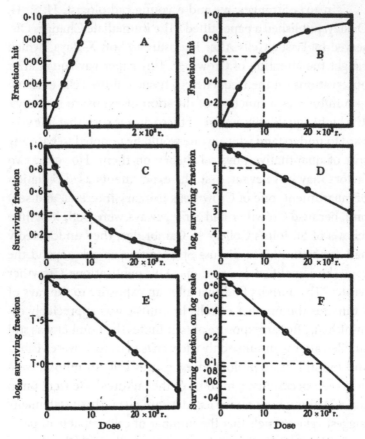

Figure 1. Target theory survival curves. One-hit target theory: 1. there is a region, the target, in which an ionization always causes a "hit" (a lethal event, a mutation, etc.); 2. hits are proportional to the dose of radiation, if the dose is sufficiently small (panel A); for larger doses, the proportion of targets that are hit saturates at higher doses and results in a convex curve tending to 100 percent (panel B); the fraction of un-hit targets (the *surviving fraction*) decreases with dose, and the decrease is exponential (panel C); on a semi-log scale, the surviving fraction decreases linearly with dose.

Simple mathematical derivation: If n_0 is the initial number of organisms and n is the number that survive after a dose of radiation, the x-rays. For this derivation he assumed that the x-rays produce their effects by making clusters of ions that must occur in the critical structure (in his words "hit") to elicit an inactivating event. From these calculations and assumptions, he estimated that the structure would be about the size of the centrosome (the organelle that serves as the microtubule-organizing center in the cell and important for mitosis). For another observed effect reported in this paper, cell disintegration, Crowther noted that a simple assumption of one hit per target was inadequate to account for the observations but that a two-hit model sufficed. A few decades later, target theory data would become familiar to biologists for its simplicity and its wide applicability (figure 1).

Over the next several years Crowther accumulated data in his own laboratory to test his approach to the target theory. One objective was to test it on known intracellular targets to verify the calculations and assumptions of the model. It is interesting to note that Crowther nearly had to share priority for

---

D, then the proportion of organisms that are *not* hit by an increment of dose dD is $-dn/n = dD/D_0$ where D_0 is the dose required to hit an average of one hit per organism. This equation integrates to give $\log_e(n/n_0) = -D/D_0$, or the log of the surviving fraction is proportional to the radiation dose. This relationship provides a simple way to estimate D_0, the dose to produce one hit per organism (when the surviving fraction is e^{-1}, or 0.37, panels E, F). Physical determination of the spatial distribution of ionizations produced by different sorts of radiation in solid and liquid materials provides a way to estimate the physical sizes of targets without knowing much about the chemical or physical nature of the target, except that they must be sensitive, in some measurable way, to ionization. (Diagram from Douglas E. Lea, *Action of Radiation on Living Cells* [Cambridge: Cambridge University Press, 1947], 73; reproduced with permission of Cambridge University Press through PLSclear)

this discovery with none other than Dr. Martin Arrowsmith, the fictional hero of the Sinclair Lewis novel, published in the same year as Crowther's paper, who made similar observations on bacteriophage:

> The results obtained from this quantitative study permit the statement that the percentage of inactivation, as measured by determination of the units of bacteriophage remaining after irradiation by gamma and beta rays of a suspension of fixed virulence, is a function of two variables, millicuries and hours. The following equation accounts quantitatively for the experimental results obtained:

$$K = \frac{\lambda \log_e \frac{u_o}{u}}{E_o(\varepsilon - \lambda t_1)}$$

(Sinclair Lewis, *Arrowsmith*, 1925, p. 406)

Other real-life scientists soon picked up on this approach to study of intracellular processes too complex to be studied biochemically or morphologically. It was particularly appealing to microbiologists because it gave a rational explanation for the exponential dose-response curves seen with heat, ultraviolet, and ionizing radiation.

Other "Hit Theories"

Another approach to an explanation of the biological effects of ionizing radiation was elaborated from Frankfurt am Main by Friedrich Dessauer (1881–1963) and his colleagues, Marietta Blau (1894–1970) and Kamillo Altenburger. Dessauer was a professor in the University Institute for the Physical Basis of Medicine in Frankfurt and an important figure in the development of the German school of radiation therapy. In his first paper, published in 1923, Dessauer reasoned—from the quantum nature of light and x-rays and the recent physical studies on the ionization of gases—that in the cell, x-rays would be expected to interact with molecules, "electrolytes, colloids and membranes" and excite them and break them into fragments. The splitting and recombination of these molecules would, he reasoned, result in the very local dissipation of energy in the cell. This energy deposition in a small volume would cause "the small regions [*Punkten*] to experience heating to a high temperature— I will call it 'point heating' [*Punktwärmen*] in what follows." Dessauer cited Wilson's cloud chamber results as well as other physical studies in his attempts to calculate the magnitude of the heating he predicted. This calculation required not only information about the quantity of energy deposited per ionization event but also an estimate of the volume over which this energy was absorbed and dissipated. He arrived at the conclusion that the heating would be at least sufficient to coagulate, and thus inactivate, enzymes inside the cell.[13]

Dessauer constructed a model of the cell in which there were sensitive "particles" (*Teilchen*), and he realized that the fraction of particles that remained intact after irradiation with a certain dose was an exponential function of the dose.[14] He then considered the case where a particle had to be hit more than once to be heated sufficiently to be inactivated. This assumption gave the single target, multi-hit survival equation showing the number of particles, M_{rn}, which have sustained less than n hits and where M is the total number of particles:

$$M_{rn} = Me^{-\sigma t} \cdot \frac{(\sigma t)^n}{n!}$$

In search of biological data with which to test his theory, Dessauer cited experiments reported in September 1921 by Francis C. Wood (1869–1951) of Columbia University, in which Wood was observing the effects of radiation on transplanted mouse tumors. Wood's dose-response data, according to Dessauer, "did not go to zero, but on the contrary showed an exponential curve."[15]

Dessauer and his colleagues clearly applied the physical concepts of target and projectile and, in a footnote to his first paper, Dessauer acknowledged a recent essay of Emil Warburg (1846–1931) in which the photochemical effect of light is discussed in somewhat similar terms. He also acknowledged the contributions of J. J. Thomson's radiation bombardment work to his thinking.[16] While the model of radiation effects on biological material envisioned by Dessauer embodied the principles borrowed from atomic physics, the point heat theory seemed aimed at better explaining the nature of the physical event that killed cells and then, from knowledge of that process, deducing the kinds of biological processes that were responsible for the sensitivity of the cells to radiation effects. Blau and Altenburger, in the second paper on this subject from Dessauer's institute, further elaborated on the possible heating to be expected, and they gave detailed, step-by-step derivations of the exponential equation from Dessauer's first paper. They discussed the interpretation of each kind of dose-response curve, which was derived with different hypotheses: single-target, single-hit versus single-target, multi-hit. In all their arguments the number of particles in the cell is treated as a knowable variable, not an unknown to be measured by the target method. The number of particles per cell was taken as more or less known from other work: "It is known that the cell has between 10^9 and 10^{10} such [protein] molecules, between 5,000 and 20,000 molecular weight."[17] In contrast to Crowther, Dessauer apparently did not realize that his concept of projectile and target might allow him to infer something more about the nature of the target itself.

By 1927 a more detailed description of the events near the primary ionization was provided by Edward U. Condon (1902–1974) and H. M. Terrill from the physics department and the Institute of Cancer Research, respectively, at Columbia University: "The quantum of x ray energy, when absorbed, is taken up by one atom of the absorbing substance and a high-speed photo-electron is liberated. This electron moves about in the neighborhood of the place of its liberation, losing energy by collisions with atoms and causing a good deal of local ionization. It is presumably the disturbing effect of this ionization on certain colloid equilibria which causes biological action, but that question is outside the realm of this paper."[18]

Condon and Terrill noted that about 30–40 electron-volts were expended to create an ion pair, so "it follows that such an absorption of one quantum of 160 KV [kilo-electron-volt] x ray is like a highly localized burst of ionic shrapnel in which about 4,000 ion pairs are liberated in less than a millionth of a cubic centimeter." They were clearly aware that the effects of ionizing radiation were dependent on the size of the volume in which the ionizations occur, and they estimated the sensitive volumes for several biological endpoints as well as showing that some data were best described by the single-hit model and other data best fit the multi-hit model.

Target Theory and Biophysics

After Crowther's seminal paper, the next most frequently cited work again came from the Cavendish laboratory; it was published in 1936 by Douglas E. Lea (1910–1947), Raymond B. Haines (1905–1943), and Charles A. Coulson (1910–1974). Titled

"The Mechanism of Bactericidal Action of Radioactive Radiations," it reviewed Crowther's derivation but added extensive
laboratory data demonstrating the exponential "probability
law" for "disinfection" of different kinds of bacteria under different conditions.[19] Interestingly, this paper was submitted to
the Royal Society by F. Gowland Hopkins (1861–1947), professor of biochemistry at Cambridge; this circumstance is perhaps
an indication of the increasing influence of this target theory
approach.

The application of the target theory concept seemed to
mark its user as something special, not quite a biologist, not
quite a biochemist. It may even be the case that adopting the
target theory approach was one of the defining characteristics
of the new discipline of biophysics. The words of the novelist
again may capture the essence of the era. Here is Sinclair Lewis
writing on Martin Arrowsmith's x-ray inactivation data: Rip-
pleton Holabird, director of the McGurk Institute, "was as much
bewildered as Tubbs would have been by the ramifications of
Martin's work. What did he think he was anyway—a bacteriologist or a biophysicist?"[20]

The target theory approach had great appeal to physicists
who were looking for interesting problems in biology during
the 1930s and 1940s. Erwin Schrödinger, Max Delbrück, Ernest Pollard (1906–1997), and Fernand Holweck were among
the well-known physicists to whom the conceptual and experimental simplicity of the target theory held great attraction. The
application of these ideas from the atomic physics of Thomson
and Rutherford to study the biology of the gene seems to have
depended in the first instance on a conjunction of events and
an individual with specific interests and knowledge—in this
case, J. A. Crowther. Its wider application would then serve the
interests and accord with the scientific style of a generation
of physicists who were developing the new field of biophysics.

While the tools of the physicists were being exploited in other directions—for example, chemical analysis and separations with electrophoresis and ultracentrifugation, and electron optics of the new electron microscope—and while the scientists deploying these tools often called themselves biophysicists, their central explanatory framework for biological problems was often not primarily molecular. For these early radiation biologists, however, their physical methods were employed because the method provided the molecular explanations that they sought. To some extent, then, it appears that the target theory first became a central, almost defining, tool of biophysics. Later, the target theory became accepted by biologists in general as a valid approach to biological structure and function, perhaps because biology as a field became more and more dependent on chemistry and physics. Indeed, almost as an example of life imitating art, the post-doctoral work of young Martin Arrowsmith in the 1920s was recapitulated in 1950 when James D. Watson, one of our current real-life icons of molecular biology, based his Ph.D. thesis on a detailed application of the target theory to the x-ray inactivation of bacteriophage.[21]

4

Heterogeneity

Physicists Doing Biology

Many scientists working in the late twentieth century would not disagree with the "meta-version" of the central dogma of molecular biology that goes something like this:

Physics → Molecular Biology → All of Biology

Because historians of molecular biology have focused mainly on the figure of Max Delbrück, who styled himself as a "physicist doing biology," there is a general impression that the role of physics in the origins of molecular biology was limited to a few, perhaps idiosyncratic, dabblers such as Delbrück, Bohr, Szilard, and Schrödinger. Even this brief list, however, is interesting: viewed from the present, Delbrück stands out as the key figure in molecular biology, but in the 1930s Bohr, Szilard, and Schrödinger were leading physicists of their generation, all major architects of modern atomic physics and nuclear physics; Delbrück was a bright young post-doctoral scientist who seemed bored with physics. As Michel Morange has pointed out, in the 1920s and 1930s the interests of many physicists encompassed biology in a very natural way. It seems Whiggish, indeed, to neglect this broad, yet heterogeneous, tradition in our admiration of the subsequent success of Delbrück in building the APG.[1]

Fresh from their heady triumphs of the quantum revolution, physicists could view life itself as ready to succumb to their insights and mathematization of the world. While the old vitalism of the nineteenth century was exorcised from public view in the twentieth century, it was by no means vanquished.[2] As we shall see, a reincarnation of this fascination with vitalism, with "life," appeared in the wake of the revolutions in physics in the 1920s and 1930s. The strong form was Bohr's "biological complementarity" ideas that were initially so interesting to Delbrück. The weaker form of this fascination was the general belief that biology was "ripe" for new discoveries so it could become a "mature" science like physics. The pioneers of the new physics from the nineteenth century pointed the way as early as 1904 with a congress at the St. Louis World's Fair on "The Unity of Knowledge" that included as speakers Wilhelm Ostwald (1853–1932), Ludwig Boltzmann (1844–1906), Ernest Rutherford, Edward Leamington Nichols (1854–1937), Paul Langevin (1872–1946), and Henri Poincaré (1854–1912).[3]

This chapter examines the consequences of this biological physicalism by considering the diversity of individuals who contributed to the phage community and by tracing the ways that biological interests shaped their work in the 1930s, 1940s, and 1950s. This group of scientists, mostly physicists and chemists, represents different generations, diverse nationalities, and distinct academic traditions, yet all developed serious interests in biological problems. This diverse group was essential in the processes of negotiation and agreement on the research framework and its set of shared commitments that defined the APG.

This chapter focuses on self-identified physicists; the case of the chemists is complicated by the growing subdivisions in both biochemistry and physical chemistry, so they will be considered, along with some renegade biologists, in the next chapter. In order, we shall consider the actors, the questions, the methods, and the answers that formed the interwar context from which phage research and much of molecular biology emerged.

What, if anything, can we identify that might represent a research framework to characterize this development in the middle third of the twentieth century? The initial version of the research framework of molecular biology had, as its *domain* of investigation, questions of biology at the fundamental level of the cell and subcellular organization. The substantive *questions* included those related to remarkable biological specificity: how are the processes of living organisms able to produce such a wide range of results that are yet so precisely and reproducibly constructed? How can heredity be so stable, yet still allow for mutations? For some physicists, too, there were questions about the energy flow in biological systems that seemed to challenge basic thermodynamic credulity. These questions of energy impinged on the vexed problems of the nature and origin of life itself. In addition to fundamental questions shared in common, the proper *methods* to approach such questions were discussed. A strong unifying belief in this community of physicists-doing-biology was the value of various versions of radiation target theory–based analyses. Indeed, it seems that it was this common belief in the target theory approach and its successful use in physics to understand atomic structure and function that gave these physicists the confidence to take on fundamental biological questions. This shared commitment to the target theory as the right approach—in spite of diverse backgrounds, nationalities, and philosophies—served as a strong unifying factor in the early versions of the research framework of molecular biology.

To see just how this was so in some specific instances, we will briefly examine the biological work of well-known physicists who turned their attention to biological problems in the 1920s, 1930s, and 1940s. In no strong sense was there a community of "physicists doing biology" or even much recognition that networks of communication and collaboration might be possible or useful. Instead, by virtue of their shared background in physics, they had been educated, indoctrinated, and socialized in ways that almost unconsciously promoted a coherent world view as they started to think about biology. We will then look closely at three key issues: what *questions* did they consider the proper domain of investigation in biology? What *methodological approaches* did they regard as appropriate in this work? And what kind of *answers* to the agreed-upon questions did they agree to accept?

The Actors: Radiation Physicists

Considered as a whole, physicists with biological interests seem to fall into two traditions, very roughly separated by the war years of the early 1940s. The pioneers of the quantum revolution such as Niels Bohr (1885–1962), the Weimar physicists Pascual Jordan (1902–1980), James Franck, Erwin Schrödinger, and Friedrich Dessauer, and the French school associated with Marie Curie all viewed biological matter and processes as fit subjects for physicists to study, but none would commit to the full-blown identification as "biologist." Some postwar physicists, such as Delbrück, Leo Szilard (1898–1964), Seymour Benzer (1921–2007), Jean-Jacques Weigle (1901–1968), and Ernest Pollard fully embraced their exclusive commitment to biological work. Some, such as Ugo Fano and Pascual Jordan, kept a foot in both ponds, so to speak; they would work on a specific biological problem for a while, often out of necessity, but revert in the long run to more familiar or productive physical studies. The distinction between pre– and post–World War II science represents not only the general trend toward specialization and fragmentation in the sciences, but also the growing success and coherence of the nascent field of molecular biology during this period.

Physics in the early twentieth century was, to a great extent, the science of radiation: light, x-rays, "natural" radioactivity, and electromagnetic fields and forces. The interactions of these forms of radiation with matter, from black-body radiation and the photo-electric effect to the scattering and diffraction of radiation by thin foils, crystals, and gratings all provided mysteries to be explained. At the same time, effects of radiation on living matter were also observed, including effects on properties unique to biology—for example, life itself. It comes as no surprise, then, that the same scientists who studied particle scattering, radioactivity, and the nature of x-rays were also interested in how radiation could cause such profound effects on biological matter, such as cell death, skin changes, cancers, and hereditary alterations.

Perhaps the most publicly visible of these early physicists was Marie Curie, whose work on the isolation and characterization of radium was household knowledge. Curie's fame resulted in the establishment in 1914 of the Radium Institute in Paris, of which she became the first director. This institute was a collaborative effort between the University of Paris and the Pasteur Institute, and it had research goals applied to the medical uses of radioactivity. While Curie headed the research arm of the Radium Institute, the physician Claude Ragaud (1870–1940) headed its applied biology laboratory. The Radium Institute would become a central locus of the new biology of radiation well into the mid-twentieth century.

In 1919 Fernand Holweck joined Curie at the Radium Institute, initially as a *préparateur* and subsequently as a laboratory director; he would play a key, if fortuitous, role in the origins of the APG.[4] Holweck exhibited a wide-ranging interest from applied engineering physics, basic physical theory, to biology and medicine. Such a range of interests and the opportunities to pursue them seemed to be more common before our era. He applied his engineering talent to perfect a molecular pump that still bears his name, to create special lamps used atop the Eiffel Tower, and, in 1926, to improve the cathode ray oscilloscope for the first television receiver. His more theoretical concerns in physics included the measurement of Planck's constant and a practical and theoretical study of the continuity of radiation between the visible and x-ray spectra.

But in addition to his work on what might be clearly classified as "physics," Holweck collaborated between 1929 and 1932 with the physician Antoine Lacassagne (1884–1971), professor at the Collège de France, to study the effects of x-rays on biological systems. They were among the first to apply the target theory developed by Crowther, a method they called *d'ultra micrométrie statique.* Holweck was later described by Salvador Luria as "renown[ed] in the field of high-vacuum physics and also an expert in radiation biology," as well as "a wonderful experimenter and inventor."[5] From late 1938 until the fall of 1940, Holweck supported Luria at the Radium Institute, where they collaborated on radiation studies of virus size. Holweck was active in the French resistance and was eventually arrested and imprisoned by the German occupation forces. He died in prison on 24 December 1941. As an experimentalist, Holweck concentrated on "what can be done?" more than "how does it work?" types of questions. This particular approach easily allowed biological problems to come within the legitimate scope of physics.

In Great Britain, too, an interest in the biology of radiation was developing. Thomson and Rutherford's scattering approach in their studies of the nature of the atom was extended by Thomson's protégé, Crowther, in his development of a simple theory of "hits and targets," which became a key method in the early days of molecular biology.

Another of Rutherford's protégés, Niels Bohr, would come to play a predominant role in the interplay of radiation physics and biology, both before and after World War II. Bohr was a particularly wide-ranging thinker, nurtured in a Danish academic environment from childhood. His father, a famous physiologist who discovered important properties of the hemoglobin molecule and its binding of oxygen (the Bohr effect), was a close friend of the positivist philosopher Harald Høffding (1843–1931), and young Niels would become a student and later a friend of Høffding. As a young man Bohr became a protégé of Rutherford in Manchester, and he revolutionized atomic physics with his proposal that atomic electron energies are quantized, a theoretical explanation for the mystery of Balmer's empirically derived formula for the spectral lines of hydrogen.

Although Bohr soon became celebrated enough to have a personal chair in theoretical physics in Copenhagen in 1916, he still was required to teach physics to students of medicine. With the rapid advances in quantum mechanics in the interwar period and the conundrum of the wave-particle duality, Bohr became famous for his philosophical principle of complementarity, the assertion that a complete account of physical reality is possible only by taking two complementary but mutually incompatible points of view (light is a wave, and light is a particle). Indicative of his broad purview, he saw another sort of complementarity in understanding the physical nature of biological life. In "Light and Life," a famous lecture to the International Congress on Light Therapy in 1932, he argued that biologists needed to recognize that their observations on intact living things were fundamentally complementary to their observations on non-living specimens in their experiments. Bohr took this occasion to expound on his views of science, realism, and epistemology, and he ventured into biological examples, perhaps to demonstrate that his ideas applied beyond the exotic realm of atomic physics. This lecture brought him to the attention of the biologists. Here was an important physicist, right up there with Einstein, thinking deeply about the hardest biological topics. Bohr, however, did not see any fundamental distinction between "physics" and "biology"; all had to yield to the same principles of understanding. He continued this Promethean outlook for his entire career, seeing no field as unworthy of his consideration and serious thought. He would continue to welcome clear thinkers with biological problems to join his institute and to support them with advice and commentary throughout his life.[6]

Weimar Germany was the center of much of the new physics and had its share of physicists whose interests in matter and energy clearly extended to biological systems. In 1920 James Franck was appointed professor of experimental physics in Göttingen, where he collaborated with Max Born (1882–1970) of the Institute for Theoretical Physics. Franck was known as a gifted teacher; he attracted and inspired an impressive group of students, including Patrick Blackett (1897–1974), Edward U. Condon, Robert Oppenheimer (1904–1967), and Eugene Rabinowitch (1901–1973), to name a few.

Franck's main field of research involved investigation of the kinetic behavior of electrons, atoms, and molecules. Together with Gustav Hertz, he studied the properties of free electrons in gases and devised what came to be called the Franck-Hertz experiment, showing that the collisions of electrons with atoms could be detected and their energetics could be measured. These measured quantized energy levels for atomic electrons directly supported the Bohr model of the atom. Franck and Hertz received the Nobel Prize in Physics for 1925.

Franck saw various links between biological problems—especially those related to radiation—and his work on energy transfer between electrons and atoms. Later he became a major authority for his work on photosynthesis where he viewed pure physics as the heart of the problem.[7] For Franck, the use of radiation as a tool to probe biological systems was an unproblematic overlap of physics and biology. In 1935 he moved to the United States. During World War II he served as director of the University of Chicago's Metallurgical Laboratory, which was the center of the Manhattan Project. After the war the Metallurgical Laboratory developed into three units at Chicago: the Institute of Nuclear Physics, the Institute of Metals, and the Institute of Radiobiology and Biophysics. The third of these institutes attests to the close relationship between the atomic physicists and their biological concerns.

Ernst Pascual Jordan was another central figure in the development of modern physics in Germany whose interests encompassed basic biological problems in a serious way. Jordan was a physics student of Max Born in Göttingen and received his Ph.D. in 1924; as a post-doctoral researcher he collaborated with Born and Werner Heisenberg (1901–1976) on the creation of matrix mechanics. This work was important to the subsequent development of Heisenberg's indeterminacy principle in 1927. Jordan authored and co-authored a series of papers in the late 1920s that constituted the foundations of quantum field theory. In the summer of 1927, Jordan joined Bohr in Copenhagen, and through the course of their discussions Jordan started to think about biological problems. In 1932 he published his views on the role of physics in biology in his article "Die Quantenmechanik und die Grundprobleme der Biologie und Psychologie."[8] Jordan suggested that life processes themselves were linked to atomic processes, in what became known as the organicist view. "From the observable state of a person at time t it is possible [if determinism holds], with the additional observation of all influences working upon him, to unequivocally calculate his state at a later time t'."[9] Jordan saw living organisms as macroscopic entities whose apparently causal behavior was directed at some fundamental level by acausal quantum events. The organism served as an "amplifier" of these quantum events.

Jordan approached his biological research from a theoretical perspective. He was often challenged because of the lack of experimental support for his ideas. With the development of the target theory with its apparent similarity to studies in experimental physics, Jordan began to interact with Zimmer and Timoféef-Ressovsky, who were using target theory ideas to study the nature of the gene in the fruit fly *Drosophila melanogaster*. In Jordan's case, he approached biological problems using the tools, approaches, and strategies that had worked so well for him in physics. After fifteen years as one of Germany's leading "biophysicists," however, he was able to set aside biological problems and return to more conventional physics, where he again made notable contributions.

The German target theorist Friedrich Dessauer represented yet another distinct and intriguing anomaly.[10] Educated as an engineer and physicist, Dessauer was fascinated by the practical aspects of x-rays from the very early days of the twentieth century. In the course of his experiments he received damaging exposures to his face, which were eventually severely debilitating. He was a successful entrepreneur, designing and

selling new medical x-ray devices before turning to academic
work and founding, in 1921, the Institute for the Physical Foundations of Medicine at the University of Frankfort, later to become the Max Planck Institute for Biophysics. His entire scientific career was devoted to the medical uses of x-rays rather
than theoretical work on the new quantum physics. A socially
concerned scientist, Dessauer was a member of the German par-
liament for the conservative Catholic Centre party from 1924
until Hitler's takeover in 1933. As an opponent of the Nazi party,
Dessauer was forced out in 1934 and moved to Istanbul, where
he was appointed professor of radiology and biophysics. After
three years in Turkey, he moved to the Physics Institute at the
University of Fribourg in Switzerland; in 1953 he returned to
Germany, where he taught philosophy of science until his death
in 1963 from the sequelae of his early radiation exposures.

As Dessauer's life shows, one cannot ignore the cataclysm
wrought on science by World War II. Not only the redirection
of research programs and the disruption of international communication and collaboration, but also the psychological im-
pacts of politics, nuclear war, and cultural chaos—all had major
effects on science and scientists. The pioneer twentieth-century
physicists seemed to relish the freedom of thought and interest, nearly unfettered by disciplinary constraints, or, as the
historian Paul Forman has suggested, only by restrictions of
conventional scientific and philosophical assumptions. However, the demands of World War I and the practical needs of
production, means of war, and the new world order moved
science in new directions.[11] The discontinuities brought on by
World War II were, of course, acutely felt by the community
of physicists. Not only were there personal dislocations caused
by the anti-Jewish policies of the Nazis, but World War II became "the physicists' war" as the technologies of war came to
rely on scientific solutions to problems such as communica-

tion theory, enemy detection, rocketry, ordnance, sanitation, and wound infections. Sometimes in the United States entire research groups were recruited into military-related research. Two major programs were the Manhattan Project, mainly at the University of Chicago and Los Alamos, New Mexico, aimed at production of nuclear explosives, and the Radiation Lab at the Massachusetts Institute of Technology (MIT) in Cambridge, aimed at enemy detection through sonar and radar. More focused, goal-oriented "big science" was on the horizon. Vertical integration of science, the idea that science could be "managed" top-down—either by the government or by private philanthropy, already under way by the Rockefeller Foundation and the Carnegie Endowment—was supported by the success of the Manhattan Project and the wartime effort to make radar and other forms of communication a reality. This model for the future of physics fit well with the perceived needs for bigger and ever more expensive machines and equipment that only rich governments and a very few philanthropists could afford. It appeared to some participants that the era of the individual scientist was waning.

An extreme example of resistance to this new order in physics was Leo Szilard, a noteworthy and important bridge between physics and biology both before and after World War II. After making some waves with his work on fundamentals of thermodynamics he left Germany to reside in England. As early as 1933 Szilard saw exciting challenges in biological research and, according to his famous recollection, he nearly became a biologist. This plan was delayed considerably, however, because of a "eureka moment" in which he conceived of the chain reaction for nuclear fission:

> I was thinking about what I should do, and I was strongly tempted to go into biology. I went to see

A.V. Hill [1886–1977] and told him about this. A.V.
Hill himself had been a physicist. He said, "Why
don't we do it this way. I'll get you a position as a
demonstrator in physiology, and then twenty-four
hours before you demonstrate, you read up these
things, and then you should have no difficulty demonstrating them the next day. In this way by teaching physiology, you would learn physiology, and it's
a good way to begin."

Szilard then explained why he did not "go into biology" until
1946.

In September 1933 I read in the newspapers a speech
by Lord Rutherford. He was quoted as saying that
he who talks about the liberation of atomic energy
on an industrial scale is talking moonshine. This
sort of set me pondering as I was walking the streets
of London, and I remember that I stopped for a red
light at the intersection of Southampton Row. As I
was waiting for the light to change and as the light
changed to green and I crossed the street, it suddenly occurred to me that if we could find an element which is split by neutrons and which would
emit *two* neutrons when it absorbed *one* neutron,
such an element, if assembled in sufficiently large
mass, could sustain a nuclear chain reaction.[12]

Although known only for his fundamental work on energy and thermodynamic theory, Szilard immediately was recognized as a physicist not limited by traditional disciplinary
boundaries, and he was quickly accepted into the inner circle
of atomic physicists. Emigrating to the United States just before

World War II, he joined Enrico Fermi at Columbia University, where they managed to build a natural uranium nuclear reactor. Realizing the potential of such a nuclear reaction for a nuclear bomb and knowing that German physicists were also thinking along these lines, Szilard persuaded Albert Einstein (1879–1955)—his former teacher, colleague, and close friend—to join him in writing the famous "Einstein-Szilard Letter" (1939) to Franklin D. Roosevelt. The letter urged the president to undertake an American effort to build such a weapon to counter the German threat as well as to secure the cooperation of the Belgian government to ensure supplies of uranium from its major source at the time, its African colony the Belgian Congo (now the Democratic Republic of the Congo). Roosevelt responded positively, and the result was the Manhattan Project. Szilard joined the Metallurgical Laboratory in Chicago as part of the Manhattan Project.

Like many physicists in the Manhattan Project who came to realize the revolutionary changes that the atomic bomb could bring to warfare, Szilard raised his concerns and warnings as time went on. In particular, like many of his colleagues, Szilard thought that military, instead of civilian, control of atomic research was a mistake. After the war, many scientists, including Szilard, would leave high energy physics research, in part because it was primarily tied to military objectives. Biology seemed to be ripe for their talents as well as their consciences.

Szilard recruited a younger colleague at the Metallurgical Lab, Aaron Novick (1919–2000), to join him in 1946 to "do biology." Both began their biological education as students of Delbrück's phage course in the summer of 1947. They initially had appointments in the postwar successor to the Metallurgical Laboratory, the Institute of Radiobiology and Biophysics at the University of Chicago, where they invented a continuous flow apparatus, dubbed the chemostat, which allowed them to make precise measurement of bacterial growth dynamics under defined environmental conditions. Such experiments, evolution in a test tube, so to speak, contributed to a clear understanding of the interplay of mutation and population biology. Szilard and Novick reduced complex biological systems to a simple physical gadget that allowed for formal mathematical analyses. Ever the optimist trying to clarify murky problems, as late as 1959 Szilard made use of the target theory in his elaborate proposal to explain aging, his "aging hit" model.[13]

Most of the postwar physicists who turned to biology did so with a commitment to biology that left no room for continued work that maintained their identity as physicists. While they still may have held academic appointments in physics departments or paid lip service as "physicists doing biology," the specializing trend of postwar science and the move of physics to "big science" meant that these new biologists were left to invent their own identity, one that often became "molecular biologist."

These former physicists were a heterogeneous bunch, hardly characterized as a group or a research community at the beginning of the postwar period. The diversity of these individuals shows the ways that biology was viewed by physicists as new methods, new tools, and new knowledge became available. Two key areas in flux were the importance of two approaches to the new biology: radiation target theory and genetic analysis.

In November 1940 Edward O. Lawrence (1901–1958) invited Ernest C. Pollard to join the MIT Radiation Laboratory to begin work on development of microwave radar. Pollard was probably representative of many young physicists whose careers were irreversibly changed by the war. He had received his education in physics at Cambridge, where his doctoral research was supervised by James Chadwick in Rutherford's group; Pollard made some of the first measurements on the radius of the nucleus. In 1933 he moved to the United States and initially took an instructorship in physics at Yale University. In his early work at Yale, Pollard constructed an accelerator for the study of nuclear reactions. He offered a course on electrical discharges in gases and took this opportunity to learn about what happens when electrons "smack into" atoms, clearly an approach patterned on his work with Rutherford and Chadwick. When he went to the MIT Radiation Laboratory, however, he did not know much about microwaves: Pollard commented, "It was judged that expertise in the area was less important than understanding the practical side of radiation physics."[14]

Pollard later assessed this wartime experience as a major influence not only on his career but on those of others:

> It certainly made use of me at the peak of my powers, in the sense of the combined presence of youth and wisdom, and set me back as a basic scientist. But it did so for many, but not all. Very importantly for me, it meant that as the war ended and the development of the atom bomb became pretty much open, I could see two important factors that would have been unclear before I went to the Radiation Laboratory. The first was that nuclear physics had now left the zone of being an operation in a university between one or two professors and some students, and entered the zone of needing big equipment and an organization to deal with it. The second was that the shadow, the deep shadow, of the development of weapons would intrude, in various ways, on work in a university.[15]

While still at MIT, Pollard organized some discussion groups to consider future directions for research. He believed that the future of nuclear physics would be in the hands of those physicists who worked on the Manhattan Project, who were privy to the recent work in that field and had access to science that was still classified as national secrets. The physicists who had worked on the radar project, on the other hand, would be at a severe disadvantage in nuclear physics, having been cut off from the field for over five years of its fast-paced development. Thus, for simple reasons of career planning, the MIT group should think about other postwar research opportunities outside nuclear physics. One such option was to take up biological studies from a physical point of view. Pollard recalled:

> It seemed that without sacrificing my skills there were two areas I could attempt to enter: Cosmology, how the universe, known then to be expanding, got that way; and some kind of biological studies involving physics. The fact that I intended to return to Yale as I was supposed to, meant that I would not find it easy to get into cosmology, which, in any event required theoretical insights that I doubted were open to my temperament. So I settled on "biophysics" and obtained the permission of the acting chairman of Physics, Leigh Page [1884–1952], to teach physics and do research on biophysics.[16]

Pollard made a systematic study of the literature on the effects of physical agents on living things: temperature, pressure, ultraviolet light, electric and magnetic fields, and ionizing radiation. In this survey he came upon a paper by Lea, Smith, Holmes, and Markham in which they used the target theory developed by Crowther and Dessauer and refined by Lea to estimate the sizes of tobacco mosaic virus and the enzyme ribonuclease by the dose-response relationship for inactivation of these entities by x-rays. Pollard recalled: "I felt considerable excitement at this. Lea used a home-made x ray machine with serious problems in estimating the dose: I had a cyclotron which delivered a measurable beam and which, moreover, lent itself to changing the speed and hence ionization distribution of the particles. Fred Forro [1924–2018] and I made a crude rig and showed that bacteria could be differentially killed by various numbers of deuterons provided by the machine. This started me on one of the major research programs of my life."[17]

Pollard's research program at Yale, first conducted within the physics department, attracted a growing number of students and post-doctoral workers, and in 1955 it became the biophysics department. His group was initially supported by the Atomic Energy Commission under the rubric of study of radiation effects on biomolecules.[18]

Pollard's approach to biology is revealed in his book *The Physics of Viruses,* published in 1953. In the foreword he emphasized the nineteenth-century interest in the physics of living systems in the work of (Julius) Robert Mayer (1814–1878), Hermann von Helmholtz (1821–1894), and John Tyndall (1820–1893), and he took it for granted that biological material was within the purview of physics. He traced the twentieth-century division between biology and physics to the excitement in the "new" physics and the success of biochemistry in biology. With respect to virus studies, Pollard noted: "As these studies began, the revolution in physics also ran its course. By 1933, when [Wendell] Stanley [1904–1971] produced paracrystals of tobacco mosaic virus, the final form of quantum mechanics had been produced by Heisenberg, Schrödinger, Born, and [Paul] Dirac [1902–1984]. The physical theories underlying these chemical processes were complete by 1935. It is not surprising, therefore, that physics began to be applied to viruses to an increasing extent from that time on."[19]

In this book Pollard treats viruses mostly as physical entities: their structure, shape, hydration, sensitivity to physical agents such as heat and radiation, and sonic and osmotic insults. A final chapter, "Virus Genetics, Virus Recombination, and Virus Physics," covers the more biological aspects of virology. This book reflects the deep underlying assumption of Pollard's view of biophysics: that understanding the physical structure of the world would eventually permit an understanding of its functional aspect, including the property called "life." His textbook *Molecular Biophysics*, written in 1962 with Richard B. Setlow (1921–2015), emphasizes this reductionistic approach.[20] Pollard's research program on using radiation effects on biomolecules seems predicated on the belief that knowledge of these components in the cell, as determined by radiation inactivation studies, would permit construction of models of the living cells, just as the analogous bombardment experiments of Rutherford and his group led to models of the atom and its nucleus.

A different research trajectory into biology is exemplified by Jean-Jacques Weigle, who was professor of experimental physics in Geneva and worked on problems of optics and electromagnetism.[21] Weigle and his collaborators studied the design of the electrostatic lenses and began to apply a Swiss-made electrostatic electron microscope to a variety of research problems. By 1950 he had set up an "electron microscopy group" and recruited Eduard Kellenberger (1920–2004), his former student, to lead this group. One application Kellenberger made was to study bacterial anatomy with the new technique.

In the late 1940s Weigle and Kellenberger decided to study the intracellular state of bacteriophages as a complement to the study of the bacterial host. This research led Weigle into contacts with other phage researchers, especially Delbrück in Pasadena. By the mid-1950s Weigle became so involved in phage research that he resigned his professorship in Geneva and accepted a research fellowship at Caltech. Still, at Caltech he continued his teaching in physics by giving a course of lectures on x-ray diffraction and crystallography for several years. In Weigle's case, his physics seemed to evolve naturally into biology, mediated by the instrumentation that was common to both.

World War II, chance, and necessity all played a role in the career of Ugo Fano, who studied physics at the University of Turin where his father was professor of mathematics. Fano arrived in the United States in 1939 as an Italian Jewish "quota emigrant." As it turned out, Fano, a scientist not usually remembered in the pantheon of phage workers or molecular biologists, would be the individual who by contingent circumstance was to be the key intersection of several strands of research and personal connections that would (1) establish the laboratory at Cold Spring Harbor, New York, as the intellectual home of the APG and (2) connect the founders of the phage group to each other and to the Cold Spring Harbor Laboratory.

Although we may wish for law-like causal connections in everyday life, we cannot overemphasize the importance of chance events in history. Contingencies are events that do not have to be the way they are; they are happenstances that may or may not have identifiable causes. In the complex world of an unknown number of causal variables, the events that captivate us are almost always "massively underdetermined," to use the language of inductive logic. Ugo Fano's role as a catalyst in the development of phage research and, ultimately, molecular biology exemplifies this important historical principle.

During his last days in Rome, Fano had been encouraged

by Fermi, after they both attended a lecture by Pascual Jordan, to take up the study of biophysics from a fundamental perspective. The potent effects of x-rays on genetic materials suggested that radiation biology might identify the microscopic nature of the gene, a view championed by Timoféef-Ressovsky, Zimmer, and Delbrück in their 1935 paper that was gaining significant note. In the 1930s a small group of scientists who were interested in the gene took their lead from this paper and saw value in the interdisciplinary collaboration among physicists, chemists, and biologists. Several small workshops were convened as "Gene Conferences" and featured eminent physicists and biologists. The "Second Gene Conference" was held in Spa, Belgium, in the fall of 1938; its participants included the physicists William T. Astbury (1898–1961), Pierre Victor Auger (1899–1993), J. Desmond Bernal (1901–1991), Pascual Jordan, Karl Zimmer, and Fano, together with the biologists Torbjörn Caspersson (1910–1997), Cyril D. Darlington (1903–1981), Boris Ephrussi (1901–1979), Hermann J. Muller, Nikolay V. Timoféef-Ressovsky, and Conrad Hal Waddington (1905–1975). Before attending this conference, Fano was able to visit Timoféef-Ressovsky in Berlin to fortify his contacts and understanding of radiation genetics. The North American version of the "Gene Conference" was held in the summer of 1940 at Woods Hole, Massachusetts, and was attended by the physical scientists Alexander Hollaender (1898–1986), Daniel Mazia (1912–1996), (?) Uber, Dorothy Wrinch (1894–1976), and Fano, together with the biologists Anton J. Carlson (1875–1956), John Gowen (1893–1967), Berwind Kaufmann (1897–1975), Karl Sax (1892–1973), Lewis Stadler (1896–1954), and Milislav Demerec (1895–1966). When Fano arrived in the United States at the end of June 1939 he already had been in contact with the radiation geneticist Timoféef-Ressovsky, who had ties to the Cold Spring

Harbor Laboratory and who was known to the established physicists interested in target theory and the nature of the gene.[22]

Milislav Demerec, director at the Cold Spring Harbor Laboratory, decided that the Carnegie Institution at Cold Spring
Harbor on Long Island in New York needed more research in
biophysics, and he was looking for a physicist for his staff. As
he noted in 1940, "Since genes and chromosome threads are
molecular structures and irradiation is effective through atomic
excitations, the study of the observed changes [mutations induced by radiation] resolves itself into an analysis in which the
knowledge of atomic physics plays a crucial role." Fano had
received help in finding a short-term position at the National
Institutes of Health with Alexander Hollaender, and Demerec
recruited Fano to spend the summer of 1940 at Cold Spring
Harbor. By the end of the summer Demerec sought funds to
keep Fano on for an extended period and even envisioned him
eventually joining the permanent staff there. Fano had attended
the Third Gene Conference at Woods Hole in 1940, and in Demerec's view, "Dr. Fano showed during discussions an unusual
grasp of biological problems and made such a good impression that [there is] a strong desire among the members of the
Conference to keep him working in our field." Fano's fortuitous association with Cold Spring Harbor Laboratory would
give that institution a direct link to the authors of the famous
Three-men-paper; moreover, his several connections to his fellow Italian refugee Salvador Luria would bring Luria into the
Cold Spring Harbor orbit as well.[23]

Soon Fano would be "borrowed" by the U.S. government
for a war-related project that needed a broadly trained physicist.
Perhaps this came about because of the dual position of Vannevar Bush (1890–1974) as chairman of the board of the Carnegie
Institution and as a key science bureaucrat for the government.

Fano worked with Demerec first on problems of radiation mutagenesis in *Drosophila,* the fruit fly, and then on similar problems in bacteria. Together they published the first description of the full set of "T-phages" (T1–T7), which became the universe of phages accepted by the APG. After the war Demerec hoped to establish a permanent group at Cold Spring Harbor to work on microbiology and bacteriophages, but Fano wished to return to physics; as Demerec noted in a letter to Delbrück, "Fano is not happy doing experimental work, and would prefer it if an arrangement could be made whereby he would spend the major part of his time with the Carnegie group in Washington [Carnegie Institute of Terrestrial Magnetism] working in theoretical physics and serving as a link between them and us." Fano did indeed return to physics: he took up a professorship at the University of Chicago, where he pursued his interests in theoretical physics for the rest of his career.[24]

This brief account of some "physicists doing biology" in the late 1930s and 1940s shows both the diversity of this community and its commonality: sometimes collegial connections, sometimes student-teacher relationships, but almost uniformly shared theoretical underpinnings in the field of physics and a belief in the new physics of radiation and quantum theory, as well as a philosophical commitment to materialist reductionism. Leo Szilard perhaps captured the attitude of his contemporaries when he said that physicists believe that "mysteries can be solved," whereas biologists rarely do.[25]

Another group that turned to biology at this time included several young scientists who had just come of age in the postwar period and were considering their pathway forward. Gunther Stent, who would later become a leader in the APG, was just finishing his doctoral work in physical chemistry at the University of Illinois in 1947. In taking stock of his personal qualities and his intellectual ambitions, he observed that these considerations had "fortified my decision to get into a new field, preferably something biological. It is not at all impossible that we are now in the 1870s of Biology, i.e., the time at which a hithertofore [sic] primarily empirical and phenomenological science is put on a more theoretical and *a priori* basis, and the prestige one brings oneself as a Ph.D. physical chemist ought to get one someplace in biological circles. In fact, I am listening in on a course in physiology this summer, just to show you that I am serious."[26] Stent finished his dissertation and received his degree in 1948 by which time he had contacted Max Delbrück to inquire about a post-doctoral position at Caltech. Following his first meeting with Delbrück in the spring of 1948, he wrote to his grad school friend Bill Treumann (1916–2016), "I guess I am going to work on bacteriophages, which sounds very interesting, though please don't ask me what that is. This summer (beginning June 28), I hope to take the phage course at Cold Spring Harbor, and then I'll be in position to enlighten you."[27] Stent went on to work with Delbrück, then with the phage group at the Pasteur Institute; he developed an extension of the radiobiological target theory as a way to understand intracellular phage synthesis.

Another young physicist who followed a similar path into the APG was Seymour Benzer. After Benzer finished his Ph.D. in solid state physics at Purdue University, he took up an assistant professorship in that department but was granted a leave of absence to explore some new areas at Oak Ridge National Laboratory. He had read about the use of the target theory by Luria and Latarjet to study intracellular phage multiplication, and he had ideas about how to refine and improve this approach. In exploring this new field, he was also delving into a rather radical new career in physics, away from the future of semiconductors and toward the future of genetics. After a year at Oak Ridge, Benzer begged his chair at Purdue for another year's leave: "I have really fallen in love with phage and want to devote a couple of years to it, at least." His plan, which he was able to realize, was to spend a year in Pasadena with Delbrück.

Not all such "conversion stories" ended as well. In 1945, Bohr recommended one of his promising young physicists to Delbrück, because he wanted "to acquaint himself with biological problems." Thorbjörn Sigurgeirsson (1917–1988) enrolled in the phage course in its first summer of 1945. With support from the Rockefeller Foundation, he secured a place in the laboratory of the Nobel Prize virologist Wendell Stanley at the Princeton Labs of the Rockefeller Institute, "to study the methods and experimental techniques used in research work on viruses." By the fall of 1945, however, Sigurgeirsson had a change of heart; his mentor in Iceland wrote to Stanley that "it seems that he wants to change his plans and go back to work on atom physics. I was very much disappointed to hear this as I thought I had made it perfectly clear to him before he accepted the position in your laboratory and the fellowship from the Rockefeller Foundation, that there was really no way back." Apparently there was. "Now that Mr. Sigurgeirsson seems to have made up his mind there is nothing I can do but to tell him that as far as I am concerned he is free to go his way."[28]

Perhaps the spirit of the times was best captured by the physicist George Gamow (1904–1968), who wrote to Wendell Stanley at the Princeton laboratories of the Rockefeller Institute in 1947:

There seems to be an epidemic among the physicists, "maladia biologica" you may call it. There are two things in this connection which I want to ask you. 1. The Office of Naval Research (in which I am a consultation in physics section) spends, as you may have heard, millions of dollars on pure research. . . . Its physics section has two subdivisions; nuclear physics and solid state. Well, the second division became interested in "aperiodic crystals" or in plain language, genes and viruses. So they want to spend money on subsidizing research in this direction on very broad lines. On the lines of physics and biology collaboration of course. The moral: may I and Mr. Mackenzi (in charge of the ONR's solid state) come to Princeton sometime the end of March to talk to you about the way to spend few hundred thousand dollars? It isn't joke, it is serious![29]

Another contingency must also be considered: a widespread, yet not universal, discomfort among physicists at the horror of nuclear war. As Robert Oppenheimer, a major architect of the atomic bomb project in the United States, noted, the postwar physicists realized that they had "known sin" as the consequences of Hiroshima and Nagasaki came to be understood. As Nicolas Rasmussen has described, for such physicists the opportunity presented by applications of physics in biology was an attractive way to compensate for the collective guilt that seemed pervasive.[30] He suggests the postwar "biophysics bubble" can be attributed to motivations toward more peaceful uses of atomic energy, such as the radiobiology of phages might offer. Another historical contingency that cannot be ignored.

The Questions

One feature that provides structure to a forming discipline is the nature of the themes and questions that begin to define its domain of study. What were these questions that seemed so seductive to some physicists in the mid-twentieth century, these inquiries that could compete with the challenges, mystery, and successes of relativity and quantum mechanics? One way or another, it seemed that "how does like beget like?" was the common question that burned for an answer. For some physicists, this question was refined to focus on the remarkable stability of heredity; unbelievably complex organisms reproduced themselves with the same complexity, down to the seeming most minute detail. Mutation rates were improbably low, and for some physicists they seemed to defy basic laws of thermodynamics. The celebrated "Three-men-paper" by Timoféef-Ressovsky, Zimmer, and Delbrück in 1934 attacked the dual and seeming incompatible observations of detectable mutation frequencies and genetic stability. Physicists saw such fundamental biological problems in general terms rather than in mechanistic terms, much like the nineteenth-century schism between thermodynamics and the detailed studies of individual chemical reactions that were, of course, subject to the generalizations of thermodynamics. As noted by Hermann von Helmholtz: "There is a sound core in this whole movement, the application of thermodynamics to chemistry. . . . But thermodynamic laws in their abstract form can only be grasped by rigorously schooled mathematicians, and are accordingly scarcely accessible to people who want to do experiments on solutions and their vapor tensions, freezing points, heats of solution, etc."[31]

As the philosopher of science David Grandy phrased it, "if physics was all precision and no promise, biology, particularly after [H. G.] Wells, was promise without precision." The hubris of physicists, fresh from their flush of successes with the quantum revolution of the early twentieth century, was noted by Léon Rosenfeld (1904–1974): "It is difficult to those who did not witness it to imagine the enthusiasm, nay presumptuousness, which filled our hearts in those days. I shall never forget the terse way a friend of mine (now a very eminent figure in the world of physics) expressed his view of our future prospects: 'In a couple of years,' he said, 'we shall have cleared up electrodynamics, another couple of years for the nuclei, and physics will be finished. We shall then turn to biology.'" Disciplinary arrogance or intellectual clarity? At least in the short run, the simplifications of biology envisioned by these physicists would yield high returns in terms of new understanding of their fundamental question, that of biological reproduction.[32]

While current discourse in biology, especially molecular biology, frames the "like begets like" problem as "replication," the initial discussions in the 1930s and 1940s were decidedly more biological and employed the word "reproduction" almost exclusively. This is of more than trivial interest because of the connotation that "replication" has in terms of both object and mechanism. "Replication" suggests exact copying, to be sure, but also refers to simple mechanical operations, basically passive, such as printing, stamping of parts, and other forms of mimicry. Reproduction, on the other hand, was something more actively done by biological objects: people, rabbits, and (maybe) microbes. It was closely connected to the concept of "growth," an essential attribute of life. The goal was to frame reproduction as a general concept, detached from messy, biological explanations. It would then fit well into a physicist's thinking about formal systems, properties, law-like behaviors, and the like. Once the discourse started to encompass "replication," one descended into Helmholtz's messy world of the chemist, confronted by molecules, specifics of chemical reactions, "heats of solution, etc." As we shall see, one of the beauties of the experimental tool provided by phage was that problems of reproduction appeared reducible to a simple "black box" problem.[33] Since all viruses appeared the same and seemed biologically simple, they could be considered similar to uniform atomic particles, as identical to each other as the electrons in free space that made progress in fundamental physics so fruitful. No more messy biological variations and individuality, the very characteristics that so intrigued classical biologists. Phage would be the hydrogen atom of biology.

The study of heredity, the remarkable property of living things, evolved during the early years of the twentieth century into the science of "genetics," a science that tacitly accepted a material basis predicated on the existence of more or less discrete objects called "genes." From the era of Gregor Mendel (1822–1884) and August Weismann (1834–1914), genes somehow gave rise to the phenomenology of heredity. This connection between the physical object and the observed outcome was a central mystery in genetics, sometimes formulated as the "genotype-phenotype" problem. Research in genetics was frustrated by a lack of understanding of the physical nature of genes: how could one study how they worked to produce the observable properties of an organism if one did not know what a gene was? One approach was to treat the gene as a formal rather than a physical object. If one could not know the chemistry and physics of genes, at least one could find out the formal rules of their behavior; much of genetic research in the first half of the twentieth century did just that. Exquisite experimental systems in fruit flies, corn, and mice were developed to study the statistical regularities of gene behavior—and the occasional aberrations of such behavior—but these investigations were conducted without any knowledge of the material basis of such behaviors.[34]

For physicists accustomed to treating atomic particles according to formal rules rather than knowledge of their underlying nature, formal genetics had a familiar ring. Indeed, geneticists talked of "the genetic structure" of a system almost the way mathematicians spoke of mathematical "structure" in abstract group theory or formal logical systems. If there was a familiar and comfortable area in biology where a physicist might find kindred souls, it was in genetics, and this certainly accounts for the preponderance of physicists-turned-biologists who dabbled in, or totally embraced, genetic research.

The Methods

Another criterion for discipline formation in the framework model is the establishment of consensus as to the nature of legitimate experimental methods likely to advance understanding of the group's research aims. In the case of the community of physicists who became interested in biology in the inter-war period, one methodological approach seemed dominant, that employing radiation as a key tool. This was not surprising for two reasons: first, radiation was a key subject of the "new physics" involving both the practical uses of x-rays and the deep theory of quantum mechanics, and second, the *éminence grise*, Niels Bohr, anointed "light" as of crucial relevance to "life" with his famous lecture in 1932.[35] Other physicists, too, believed that just as the study of light and other forms of radiation had proved so fruitful for twentieth-century physics, so, too, the study of the interaction of radiation with biological systems would unfold the mysteries of life. Some, like James Franck, saw a direct connection in the process of photosynthesis, and others, such as James Crowther, Friedrich Dessauer, Pascual Jordan, and, eventually, Max Delbrück, would see high energy radiations as an atomic tool to probe cells at a level of resolution not yet possible with the existing techniques of morphology and chemistry.

If the major questions for the new biologists were about genes and reproduction, their methods were selected to match the imagined size scale of the cell machinery involved. Sub-microscopic structures and molecules would require chemistry and physics of the same scale. Radiation, with waves and particles on the molecular scale, was a tool they understood. Clear endpoints for a "hit" were essential. Assays should be quantal, the outcome should be yes or no, a function should be present or absent, an organism should be fully living or otherwise dead. Methods that gave results that were simply a measured value—the size of an organism, its life span, the amount of sugar in its liver, all observations that represented the totality of the microscopic state of an organism, its molecular average so to speak—would not allow interpretation at the fundamental level that these physicists sought. Even for complex systems such as an intact cell, the target theory with its all-or-none, life-or-death "hits" by a lethal particle of ionizing radiation could suggest the size and shape of the critical intracellular object that was required for life. If the endpoint was not life or death, but a specific, observable mutation, say, eye color in the fruit fly, it might be possible to "see" a gene by radiation target theory experiments even though one did not understand the proximate causes of eye coloring in the least.

While radiobiology as a method had wide appeal, at least in the beginning, genetic methods were unfamiliar to the early physicists turned biologists. Consensus on the proper biological material—plants, animals, microbe... invertebrates, insects, fungi—was harder to develop and, indeed, did not emerge. One's choice of an experimental organism turns out to have complex psychological aspects involving taste and temperament as well as practical aspects such as cost, laboratory facilities, and availability. Plant genetics, the favorite approach of the classical biologists, was too qualitative and too slow for scientists looking to eliminate ambiguity and uncontrolled variables. Microbes, which seemed simple at the time, which grew incredibly fast, and which took very little time to master in the laboratory, were the obvious choice for some. Bacteriophage, the simplest of microbes, were best of all, but were they complex enough to exhibit the interesting properties of life that were worth studying? By the time Delbrück, Luria, Hershey, and their few acolytes settled on phage, they knew that at least phage could reproduce, and reproduce faithfully: a specific phage, characterized by plaque morphology, by host strain specificity, and by antigenic properties, would always retain these properties over many cycles of reproduction. Soon they observed that rare variant phages occurred and that these variants were stable, indeed, seemed to be "mutants" just like in higher organisms. Genetic studies with phages might be possible. In what may have been a leap of faith, or possibly desperation, phages became the methodological organism of choice, central to the framework of shared commitments of the APG.

The Answers

Answers are not generally known beforehand, but what will count as an answer, what form an answer will take, and what sort of evidence will be accepted to support belief in an answer are all knowable (actually, agreed upon by consensus rather than truly known) in advance. Indeed, these are all notions that might well belong in the framework of shared commitments of a nucleating research community. Delbrück, especially, was seen as concerned with these matters. But he was not the only one. In the 1930s, Max Mason (1877–1961) and Warren Weaver (1894–1978) at the Rockefeller Institute mapped out a program that envisioned a biology based on molecular answers instead of organismal answers. They used the financial resources of the Rockefeller Foundation to nudge (force?) biological research toward what they saw as critical problems of molecular structure and function, much the way physicists focused on the physical world at the level of the fundamental building blocks of nature, a program that in 1938 Weaver called "molecular biology."[36]

One of the recurrent themes from memories of phage course graduates is Delbrück's emphasis on good experimental design as necessary for good answers. Simple theories, first-order theories, were favored over complex or vague ones.[37] Whereas many biologists avoided or distrusted mathematics, the APG, more trusting of the discipline, embraced mathematics when it seemed useful.

Conclusions

These brief vignettes illustrate the diverse yet common nature of the interface between physics and biology in the first half of the twentieth century. This interface encompasses the deeply philosophical concerns of Bohr and Jordan; the eclectic interests in the natural world exemplified by Szilard; the extension of prior perspectives to frame new problems, as seen in the work of Franck, Pollard, and Weigle; and the opportunistic forays into biology of Holweck and Fano.

These scientists represent a much larger group of individuals, trained in physics and often highly successful in that field, who saw biological research as within their natural domain, as a legitimate subject for physical investigation using the approaches, tools, and modes of thought common to physics. Before the war, individual physicists indulged their curiosity and could explore biological phenomena related to their physical interests, often connected with radiation. After the war, however, to continue in physics often meant a commitment to more bureaucratic, less individualistic science, and, for some, association with the guilt of nuclear warfare. Those who would focus on biology were often forced to choose between the ever specializing and diverging worlds of biology and physics. The diversity that initially characterized the early physicists doing biology began to diminish by the 1950s, however, as consensus began to emerge as to key questions, accepted methodologies, and standardized materials. This consensus would begin to provide the research framework upon which the new discipline of molecular biology would be built.

5

Heterogeneity
Chemists and Biologists

One of the persistent myths about the APG is that it was a group of "physicists doing biology," a notion propagated by Delbrück on multiple occasions. This characterization, like most myths, has a kernel of truth, but any discussion of the APG would be remiss if it did not consider scientists who identified as either chemists or biologists. The formation of the APG and the development of its own research framework required incorporating commitments from chemists and biologists as well as physicists. These three research traditions, disciplines, or schools of thought had distinct, sometimes conflicting ideas about phage research, yet in the end, these ideas were successfully molded into the framework of shared concepts that unified and defined the APG. Indeed, physics, chemistry, and biology would not be considered distinguishable "disciplines" were it not for equally distinguishable research frameworks. For the APG founders, a significant task was to meld, adjust, and negotiate research commitments held by existing disciplines into a communal sharing of a new set of phage-oriented research commitments.

The previous chapter described the heterogeneity of physicists who adopted phage research; this chapter will extend this study to the even more heterogeneous groups of chemists and biologists who joined the APG.

The Biochemists

Widely recounted in the lore of the APG is Delbrück's disdain for biochemists, yet biochemists and biochemical experiments were ever present in the APG from the beginning, and, indeed, it was biochemistry that eventually ushered phage genetics into the new Elysium of molecular biology. Was this just a controversy over professional vanity and hegemony, or was something else at stake? What are we to make of this persistent part of our origin story?[1]

In the fall of 1949, Delbrück argued the case against biochemistry in a lecture to the Connecticut Academy of Arts and Sciences:

He [the physicist] may be told that the only real access of atomic physics to biology is through biochemistry. Listening to the story of modern biochemistry he might become persuaded that the cell is a sack full of enzymes acting on substrates converting them through various intermediate stages either into cell substance or into waste products. The enzymes must be situated in their proper strategic positions to perform their duties in a well-regulated fashion. They in turn must be synthesized and must be brought into position by manoeuvers which are not yet understood but which, at first sight at least, do not necessarily seem to differ in nature from the rest of biochemistry. Indeed, the vista of the biochemist is one with an infinite horizon. And yet, this program of explaining the simple through the complex smacks suspiciously of the program of explaining atoms in terms of complex mechanical models. It looks sane until the paradoxes crop up and come into sharper focus. In biology we are not yet at the point where we are presented with clear paradoxes and this will not happen until the analysis of the behavior of living cells has been carried into far greater detail.[2]

Often this lecture, reprinted in *Phage and the Origins of Molecular Biology*, has been read as Delbrück-the-physicist's Olympian view of chemistry as mere "stamp collecting," to use another common pejorative phrase of the time. It can, however, be read as his polemic challenge, and a hope for help from biochemists who might join with the physicists to provide the details of what might be inside the "black box" posited by the target theorists. This reading is amply supported by the fact that Delbrück actively recruited hard-core biochemists into his research group; encouraged his protégés, such as Jim Watson, to study under biochemists such as Herman Kalckar (1908–1991); and, albeit rather clumsily, engaged in biochemical approaches to phage problems that intrigued him.

Delbrück's little phage club was riddled with biochemists, scientists less enamored by the mysteries of the gene and its metaphysics as by the challenges of macromolecular synthesis, an outgrowth of the realizations of metabolic biochemistry. The cell was not just humming along with day-to-day wear and tear and the needed repair but was making macromolecules according to specific chemical processes and in response to physiological requirements. They saw phage, with their simplicity of composition, their rapid and amenable growth cycles, and their amenability to the new isotopic methods, as the ideal system to understand finally the biosynthesis of macromolecules, initially proteins and eventually nucleic acids as well. The chemists and the physicists started with different grails, but in phage, they found common clues.

That Delbrück's view of biochemistry might have been disingenuous is suggested by his deliberate recruitment of young biochemists into his orbit. Indeed, it seems that his exertions in attracting biochemical types eclipsed his efforts to collect physicists. In his negotiation with Caltech for his move there from Vanderbilt, he specifically noted that he would need resources to bring a biochemist into his research group. He almost immediately recruited Wolfhard Weidel (1916–1964), a young German protégé of both Adolf Butenandt (1903–1995), the biochemist's biochemist, and Georg Melchers (1906–1997). Gunther Stent described Weidel's challenge to Delbrück's group during a lab seminar: "Well, gentlemen, once you fabulists are done dreaming up your fantasies, let a biochemist tell you how things are in the real world."[3] The word "biochemist" soon became a lab term of derision.

Very soon after his move to Caltech, Delbrück made contact with another young biochemist from New York University, Mark Adams (1912–1956), who would become Delbrück's alter ego for the first decade of the phage course at Cold Spring Harbor. Adams received his Ph.D. (1938) in chemistry at Columbia, working on the enzymology of tyrosinase with the venerable organic chemist John M. Nelson (1876–1965), teacher of John Northrop (1891–1997), Roger Herriot (1909–1992), and Joseph Fruton (1912–2007), among other notable biochemists. After his Ph.D., Adams spent two years with Walther F. Goebel (1899–1993) at the Rockefeller Institute studying the chemistry of polysaccharides. When it came time for the newly minted doctor Jim Watson to expand his horizons, Delbrück and Luria

advised him to learn some biochemistry and recommended the
biochemical group in Copenhagen headed by Herman Kalckar.
Kalckar seemed to be Delbrück's sort of biochemist. They had
developed a personal and intellectual relationship before World
War II while they were both visitors at Caltech, and both seemed
to have an eclectic view of science. Delbrück convinced the
biochemist Kalckar to take the famous microbiology course at
Pacific Grove, California, taught by C. B. van Neil (1897–1985),
an event that Kalckar later described as life-changing. Kalckar
became, along with Fritz Lipmann (1899–1986), the authority
on phosphate metabolism, and phosphate, in the form of nucleotides, was central to DNA chemistry. Perhaps that was the
reason Delbrück thought Watson should learn some phosphate
biochemistry.

No doubt there were traditional disciplinary boundaries
involved: the education, discipline-wide fundamental assumptions, and evidentiary standards differed significantly between
most physicists and most biochemists. Certainly, mid-century
biochemists disagreed as to whether they were chemists who
studied chemistry of biological material or biologists who used
chemical approaches to traditional biological problems. Department naming reflected this dispute; biological chemistry
versus biochemistry suggested subtle differences in one's intellectual commitments. Where did organic chemistry end and
biochemistry begin?

Education in physics and biological and organic chemistry differed significantly. Physics in the early to mid-twentieth
century was flush with success at reductionist simplification with
the advent of the quantum revolution. Along with nineteenth-
century thermodynamics, quantum physics was providing the
broad, general "laws of nature" that reduced the universe to
a few simply stated (but ultimately very mysterious) principles
from which a myriad of specific cases could be derived. One

did not need to master an encyclopedia of individual observations to be a physicist.[4] Organic and biochemistry, on the other hand, was exploding with new observations and new methods that had yet to be reduced to a few general laws. Learning and remembering a compendium of eponymous reactions, structural formulae, processes, rules and exceptions made up the education of most organic and biological chemists. It's not surprising, then, that scientists who were educated and socialized in these two disciplines should view their work in different ways. Yet for a small group from each of these existing disciplines, phage provided a fascinating nexus that would serve each in similar ways.

In his always succinct way, Jim Watson described "a sharp gap" during his summer at Cold Spring Harbor in 1948 between the biochemist (Seymour Cohen [1917–2018]) and the physicists (Delbrück and Luria): "Cohen wanted biochemistry to explain genes, while Luria and Delbrück opted for a combination of genetics and physics."[5] Did this tension extend to the framework of shared commitments and threaten the development of the APG? Were such tensions resolved and if so, how? I believe these tensions were real, the legacy of ingrained disciplinary attitudes, but in the long run they had little lasting consequence. Scientists from both traditions shared basic commitments with respect to fundamental problems and openness to new approaches necessary for phage research. The ongoing revision of those commitments based on new knowledge and on new methods was both natural and essential to the development of the APG into a key contributor to the nascent field of molecular biology.

It is now important to review in more detail the discipline of biochemistry in the immediate pre–World War II period— that is, the community from which the phage biochemists emerged and the way their thinking influenced the research

framework of the APG. Biochemistry and its sibling, physiological chemistry, were fundamentally subdivisions of chemistry, organic chemistry in particular. The leaders of biochemistry
in the late 1930s were educated in organic chemistry, and they
focused on the molecules from biological materials. Academic
departments debated their names and identities: physiological
chemistry, biological chemistry, or biochemistry—all different
flavors of chemistry. Organic chemistry involved determining
the structures and reactions of organic molecules, devising ways
to synthesize them in the laboratory to provide cheap and abun-
dant sources of useful natural products: dyes such as indigo
for blue jeans, small molecules for drugs such as aspirin and
sulfa, and vitamins for nutritional supplementation. Students
of organic chemistry, even into the 1950s, became experts on
the myriad reactions that could be used to synthesize organic
molecules from basic starting materials. Exams at that time
often took the form of "roadmap problems": propose a syn-
thetic pathway for compound X, starting with A, B, and C.
Physiological chemistry, often the rubric of biochemistry in
medical schools, was devoted to the organic chemistry of compounds useful to medicine: blood glucose, metabolic reactions
of foodstuffs related to nutrition, the chemical makeup of normal and pathological body fluids, and the like.[6] Comparing
texts on organic chemistry with those on biochemistry in the
1940s and 1950s reveals a striking similarity, both with emphasis on structure and reactions of small molecules.

However, it is an oversimplification to ignore the differentiation of the field of chemistry into distinct subdivisions,
based on physics, that started at least in the late nineteenth
century. As historian John W. Servos described in his master-
ful study of physical chemistry, some chemists were distinctly
physicalists, interested in molecules as physical bodies, but
most chemists focused on the reactions and changes involving

small molecules, with the carbon-containing organic compounds providing a dazzling array of opportunities. The latter part of the nineteenth century saw the beginnings of a fundamental unification of chemistry and physics that would carry through and underlie the study of phages in the twentieth century, the thermodynamics of J. W. Gibbs (1839–1902) and Hermann von Helmholz, and its statistical interpretation first by Ludwig Bolzmann and later its elaboration as molecular theories by James Clerk Maxwell (1839–1879) and others. This work formed the basis for what would become chemical thermodynamics, linking the physicists' concepts of heat and work and the chemists' experiments on chemical reactions. Physical chemistry, then, was a kind of hybrid, applying the tools of pure physics to the messy world of chemistry.

Nothing at the time was messier than the chemistry of substances that would not yield to the chemists' tools of purification, crystallization, precise elemental analysis, and characterization by so-called functional group studies (the diagnostic reactions of amino groups, carboxyl groups, and so forth). These substances, termed "colloids" in opposition to "crystalloids," were what are now called macromolecules, or simply large molecules. Although they frustrated the classical organic chemists, colloids attracted the attention of some of the physical chemists because—large and intractable as they seemed— they still could be studied as physical particles: sizes, diffusion rates, shapes, light-scattering and viscosity properties could be discerned. The crucial defining relationship between physical chemistry and biochemistry, one that differentiates biochemistry from simply the organic chemistry of biological systems, is in the science of macromolecules. Living beings are composed of many types of macromolecules, originally called "albuminoids," a category which includes proteins, nucleic acids, and polysaccharides, to name the major types. Classical organic chemistry offered little in the way of techniques to study such large and complex molecules, so chemists, *bio*-chemists, turned to methods of physical chemistry, methods based less on chemical principles than on the basic physics of small objects. Phages, as it turned out, fell neatly into this category of substances. Indeed, Félix d'Herelle, the discover of phage, viewed phage as a "living colloidal micell."[7]

By the 1940s, this earlier distinction between crystalloids and colloids had been largely abandoned, and colloids had become macromolecules in recognition that proteins and polysaccharides were indeed discrete molecules with specific covalent bonding rather than physical aggregates of smaller true molecules. Still, proteins and polysaccharides, and very occasionally nucleic acids, were treated cursorily, except as sources of breakdown products in the study of nutrition. A few intrepid chemists were attempting to study the structures of these large molecules, but there were precious few who even ventured a suggestion as to how they were synthesized, either in the living cell or in the laboratory. To a significant extent, Delbrück's view was accurate; biochemistry of the organic chemist, triumphant in mapping metabolic pathways, seemed to be mired in the arcana of organic chemistry. There were, however, glimmers of change: in 1939 a leading structural biochemist of the day, William T. Astbury, noted: "Biology is fast becoming a molecular science, a desire to tread as far as possible the friendly ground of physics and chemistry and see where it leads. It may be that the angels are right, but it is good to feel and take part in a foolishness that is the scientific hall-mark of our times. The search is now for the structure and arrangement of the molecules of living things. Chief among these molecules are the proteins, and the greatest excitement these days is about the proteins." The tentative understanding of protein structure facing the first phage researchers was captured by

Max S. Dunn (1895–1976) in 1941: "It seems probable that proteins are intramolecular, three-dimensional systems of polypeptide chains held together in folded, cyclic, or cage-like structures by sidechain bridges of hydrogen bonds. The individual chains are considered to be made up of structurally related amino acid residues which occur in definite number and sequence. Complete details are lacking concerning the precise composition and structure of any protein, but all proteins are thought to be constructed of amino acid residues and polypeptide chains bound together according to similar patterns." Dorothy Wrinch, an intrepid mathematician, tried her hand at theorizing about the structure of proteins based on geometric considerations with input from basic chemistry to guide her ideas. Some biochemists took her analysis seriously.[8]

For biochemists who had grown up as classical organic chemists, such as Hans Clarke (1887–1972) at Columbia, Wendell Stanley at the Rockefeller Institute, and Alexander Todd (1907–1997) at Cambridge, organic chemistry was the lens through which they viewed biology. Macromolecules were studied by "analysis"—breaking them down into their parts and studying the building blocks. As to function and synthesis of these large biochemicals, there was little known and seemingly less interest. Macromolecules were known as structural components of life: polysaccharides such as cellulose allowed trees to stand; proteins provided the covering and skeletons of visible cell structures such as chromosomes and sometimes acted in vaguely understood ways as biological catalysts known as enzymes; proteins in solution made up the liquid component of blood, giving it viscosity and osmotic properties; and nucleic acids were simply mysterious. Although the classical biochemists had worked out many of the fates and reactions of the degradation of these macromolecules, they had no clue as to how they might be synthesized. As Astbury observed: "The problem of protein synthesis is not one of proteins alone, but of proteins plus other molecules—saccharides, nucleic acids, etc. This ghost of a generalisation that is looming up, that different amino acid constitutions may be associated with similar structures, hints at a world behind it, and activities of which we are unaware. When proteins are born, other molecules assist at their birth; and perhaps chief among them are the nucleic acids. The earliest reproductive processes that we know, those of the viruses and the chromosomes, always involve protein and nucleic acid."[9]

But a new generation of biochemists, more versed in physical chemistry than classical organic chemistry, began to think about this problem of macromolecular biosynthesis, not only in terms of the chemistry of the assembly, the making of covalent bonds, but of the ways that precise compositions, spatial arrangements, and sequences of the components might be determined. As Astbury presciently noted in his 1939 review, viruses and chromosomes might be a good place to start.

The problem of protein synthesis, then, was of growing concern to some biochemists, especially those whose attachments to classical organic chemistry were beginning to weaken. Physical chemistry, especially related to macromolecules, did not rely on the organic chemists' traditional tools and aims. Characterization of macromolecules did not involve melting points, differential solubilities, or diagnostic reactions such as the preparation of certain standard derivatives, for example, the Hinsberg reaction using benzenesulfonyl chloride to distinguish primary, secondary, and tertiary amines. New methods of physical chemistry seemed more suited to these large structures. Sizes, that is, molecular weights, were measured not by atomic composition but by sedimentation and diffusion rates; shapes and more detailed structural information were attacked by light scattering, x-ray diffraction, and macroscopic solution properties such as viscosity. The most challenging problem, however, was how such large molecules could be made in the first place. Since protein synthesis was a cellular process, biology was a necessary part of the problem. This aspect of biochemistry was not far from Bohr's conundrum in *Light and Life:* how to study living processes in non-living conditions. To these biochemists, viruses—most likely partly or entirely composed of protein—might, when they reproduced, represent the synthesis of new protein. Virus, including phage, reproduction would be a process leading to understanding of protein synthesis.

Wendell Stanley, a protégé of the doyen of American organic chemistry, Roger Adams (1889–1971) at the University of Illinois, was famous for his crystallization of several viruses, including polio and tobacco mosaic virus. His program saw crystallization as the organic chemists' proof of chemical purity, and along the way, because the virus had biological activity, Stanley was able to extend the prior work of Sumner and Northrop on the less dramatic protein molecules that merely exhibited enzyme activity.[10]

Stanley was sympathetic to his protégé, Seymour Cohen, who saw how phage might be studied to illuminate the problem of protein synthesis. Cohen wrote from Paris in 1948:

Phage work is at an impasse. I could see DNA and protein being synthesized but mechanisms of nucleic acid and protein synthesis are totally unknown. I could amuse myself with various phage phenomena or demonstrate similar phenomena with animal viruses or go after the enzymatic mechanisms of nucleic acid formation. Well, after due consideration, I shall stress the latter but keep going on various aspects of the phage system. The two have to go hand in hand, for the present anyway, because T2-infected cells are the first system where you can have DNA synthesis in the total absence of RNA synthesis, and that's an important tool.[11]

In addition to Seymour Cohen, the young scion of the Rockefeller virus group based in Princeton and led by Stanley and John H. Northrop, Lloyd Kozloff (1923–2012) became another leading biochemical presence associated with the APG. Kozloff was a protégé of Earl A. Evans, Jr. (1910–1999); he received his Ph.D. under Evans at the University of Chicago in 1949 and would go on to be an influential phage researcher into the 1970s. Evans had received his Ph.D. in Clarke's department at Columbia in 1936, and he collaborated with David Rittenberg (1906–1970) and Rudolf Schoenheimer (1898–1941), both biochemists engaged in using the new isotopic methods to study biosynthetic pathways.[12] At Chicago Evans pioneered the study of carbohydrates with the newly available isotope of carbon, ^{11}C, showing that cells made carbohydrates from carbon dioxide. Like Cohen, Earl Evans was a biochemist taking on the challenge of macromolecular synthesis. With Kozloff, he started the study of the biosynthesis of both nucleic acids and proteins using isotopic tracer methods. Kozloff and his Chicago colleague Frank Putnam (1917–2006) saw phages as a useful system to investigate protein synthesis (initially) and later DNA synthesis.

As Kozloff observed:

Although little is known about the actual mode of virus reproduction the autocatalytic duplication of viruses is generally believed to be performed by mechanisms analogous to those functioning in the synthesis of normal cellular nucleoproteins. The

controlling influence of nucleic acids in the synthesis of self-duplicating units has been emphasized
(Caspersson). The same factors are thought to be
significant in general protein synthesis (Caspersson;
Spiegelman and Kamen). These basic similarities
suggest that information about general protein synthesis might be gained by studying the mechanism
of virus reproduction.

His enthusiasm was, however, tempered with realism in his
dissertation summary: "The complexity of the bacteriophage
is evidenced by the size, morphology, metabolic activity, and
chemical and genetical composition. These characteristics and
particularly the presence of several genes imply that there is
more than one type of protein in the phage. The study of virus
reproduction, then, as an approach to the general problem of
protein synthesis can hardly be as straightforward as originally
conceived."[13]

Kozloff, Putnam, and Evans had access to the new radioisotope of phosphorus, ^{32}P, in the form of inorganic phosphate
made at Oak Ridge, and they used this material to prepare
radiolabeled phage for experiments designed to find out the
fate of the parental phage material. They found that there was
a physical continuity in phage reproduction with about half of
the parental label being recovered in the progeny phage. This
experiment, involving transfer of radioactive parental phage
material to progeny phage particles, became a standard method
in biochemical phage research, somewhat akin to the Luria-
Latarjet protocol in target theory research. Seymour Cohen, as
well, pioneered this approach, and he found that much of the
physical substance of the phage (protein as well as nucleic acid)
was derived from the soluble medium components, not by incorporation of pre-existing cell proteins and nucleic acids. And,

of course, a later variation of this experiment was made famous by Alfred Hershey and Martha Chase (1927–2003) when they were able to test simultaneously both the protein and nucleic acid components of the parental virus for entry into host cells upon infection.

The Biologists

As with chemistry, biology too was a discipline with diverse subgroups. Nineteenth-century biology saw the beginning bifurcation of the laboratory from the field: experimentation versus natural history, exemplary models versus generalizations from diversity, physiology (function) versus morphology (structure), analysis and mechanism versus classification and description. Likewise, one's taste for plants or animals divided botanists from zoologists, and size divided microbiologists from organismal biologists. And, importantly, in the case of phage, medical and biological viewpoints were to become increasingly divisive.

   With the increasing successes of the germ theories of disease, together with the understanding of microbial life on earth, the subfield of microbiology became almost totally a laboratory-based science. Only recently has microbial ecology and evolution taken microbiology in new directions. With microbiology in the lab, it developed relationships to biochemistry as one of the major methodological ways of studying (and exploiting) microbes. Microbiologists, then, had research framework assumptions that fit well with the physicists and biochemists interested in phage. Of course, phages were originally a discovery of the discipline of microbiology. Still, most phage workers in the mid-twentieth century were not particularly sympathetic to the APG. Raettig's bibliography lists thousands of publications on phage during this period, but most were related to medical aims such as phage therapy and phage typing of bacteria rather than devoted to fundamental problems of genes and reproduction. Neither their methods nor research aims overlapped with the research framework developed by the APG.

There was, however, a coterie of biologists (including physicians) who saw the puzzle of reproduction and gene function as a worthy goal and brought their biological viewpoints to the APG. Some, such as Renato Dulbecco (1914–2012), John Cairns, and Salvador Luria, had been educated as physicians but found the laboratory more congenial than the clinic. There were other biologists who did not do phage work but nonetheless had set the stage, through their early twentieth-century advocacy, for some of the commitments of the APG's research framework; these forerunners included Jacques Loeb (1859–1924) with his writings on the "mechanistic conception of life"; Joseph Needham (1900–1995), writing on "chemical embryology"; both Thomas Hunt Morgan (1866–1945) and Hermann J. Muller, with their emphasis on model laboratory study of heredity in the fruit fly; Marjory Stephenson (1885–1948), who pioneered the study of microbial biochemistry; and more generally, such scientific leaders as J. Howard Brown (1884–1956) of Johns Hopkins University, who, in his presidential address of 1931 to the Society of American Bacteriologists, argued for more attention to the biological, as opposed to the medical, aspects of bacteriology.[14]

By the 1930s, too, some biologists had become committed to looking to physics and chemistry for new approaches to intractable old problems, such as Muller's appeal to the Soviet physicists in 1936, cited earlier, calling on the physicist and the chemist to "step in" on the problem of the gene.[15]

Conclusions

While both chemists and biologists had distinctive disciplinary research frameworks, there were subgroups of each with common interests built around questions of macromolecular synthesis that could not be easily addressed by existing research frameworks. But these questions could be incorporated, along with the physicist's questions of gene reproduction, into a shared framework of inquiry for the APG once it was accepted that the physical nature of the gene would provide a unifying answer. As the architect Louis Sullivan had pronounced about his work in the late nineteenth century: form follows function.

6

Nucleation

Formation of the American Phage Group

The American Phage Group is by now a well established notion in the history of molecular biology, yet its definition, origin, and structure remain elusive. What was it? How did it start? What did it do? These are critical historical questions that we will now address. As we have seen, diverse groups of successful physicists thought they had an entry into biology, an area they viewed as ripe for their fundamental insights from their more mature science. But how did a few (certainly not most) of them come together to form a nucleus of collaboration that would attract others and eventually create a coherent research community that would lead to the full-fledged birth of molecular biology? How did they establish the norms of this new endeavor? How did they recruit converts to the specific views of their research program? How did they deal with challenges from inside and outside the community? How was "membership" in this research community decided? Who is in and who is out? These questions are, of course, general to the development of many scientific disciplines. In the case of the group of phage researchers and their role in the development of molecular biology, the historical record is particularly rich, and some of the participants were overtly aware, at the time, of their efforts to shape the evolution of this new community and its institutional forms. What was the APG research framework and how was it developed?

What Is a "Discipline"?

Historians of science have used the concept of a "school" as a group of scientists related by virtue of their having studied with a common teacher (or the protégés of that teacher) and of having adopted the theories, approaches, or philosophies of the teacher or a local group of teachers. This way of defining a group is genealogical and is exemplified by "the Oxford Physiologists," the "Michael Foster School," and the "Delft School of microbiologists." Other schemes, related to Derek de Solla Price's idea of "invisible colleges," use networks of communication and collaboration, including co-authorship of scientific papers and a concord of citations, as ways to identify research groupings. Indeed, mathematicians quantitate such a system by the proximity to a famous and highly productive individual, Paul Erdős (1913–1996), with a number that expresses the closeness of co-authorship: one's Erdős number is 1 if one has co-authored a paper with Erdős, 2 if one has co-authored a paper with someone whose Erdős number is 1, and so on.[1]

Beyond the notion of schools, the concept of "discipline formation" is one that historians, sociologists, and even scientists have struggled with. Several approaches are possible. One way is to focus on the "actors" or "agents" and their interactions, networks, and social structures; Nicholas Mullins exemplified this approach to molecular biology in a well-known paper that documented the linkages between scientists based on joint publication and shared mentors. Another approach is to pay attention to the behaviors and activities of already established disciplines as they fragment and rearrange themselves. This method has been used to understand the rise of psychology as a reassortment of some practitioners of philosophy melding with physiologists in the nineteenth century. A third approach rests on the notion of shared theoretical and experimental commitments. We will take this last approach, drawing on the "research framework model" proposed by Barbara Von Eckardt in her study of the development of the new discipline of cognitive science.[2]

In the case of the APG, one defining characteristic was a commitment to a particular experimental organism, bacteriophage. Besides that, however, were several other crucial relationships that went beyond simply studying a particular kind of microbe. These relationships were both intellectual and social. The past tense is appropriate here because it is clear that the APG no longer exists. In its general form, it appears to have coalesced in the mid-1940s but it had lost most of its coherence by the mid-1960s. To summarize the core research commitments of the nascent APG, the *domain element* focused on reproduction, especially the nature of the gene and its remarkable stability; the *questions element* involved the nature of the gene and how it reproduces with high fidelity, yet allows rare but stable mutation; the *answers element* restricted answers to physical principles based on chemistry and physics of cell components; and lastly, the *methodology element* required simple experiments, with observable and quantitative outcomes, amenable to theoretical analysis.

Another defining feature of the APG was that it was distinctly non-medical in its outlook. This point cannot be overlooked in any aspect of the history of molecular biology. The tension between the "art of medicine" and the "science of biology" was always in the background of this story. Phage research, early on, bifurcated along these traditional fault lines: medical and biological. Physicists, even those with medical backgrounds, adopted the shared viewpoint that phages were of intrinsic interest as tools for understanding basic biological problems, not for their medical utility, conceptually or therapeutically. This attitude served as a strong boundary between phage research of the sort adopted by the APG and the work of those who employed phages for clinical diagnosis (that is, phage typing) and for therapeutic applications prior to the antibiotic era of post–World War II.

As we have seen, research on bacteriophage was taking place both in the United States and in Europe since d'Herelle's discovery of phage in 1917. Some of this research, of course, focused on the potential therapeutic applications of phage, but studies on the fundamental nature of phage were by no means neglected. By 1940, bacteriophages had become widely accepted as some kind of (filterable) virus. Thus, phages were as mysterious as other viruses and probably benefited from interest in the nature of these "living chemicals."

A key biological problem that was emerging in many areas of biology in the early twentieth century was that of the mechanisms of faithful reproduction. Cell division, chromosome duplication, and the mechanics of heredity were general problems for many biologists. Since the detailed study of heredity in model organisms such as the fruit fly and mouse, as well as "higher" microbes such as molds, yeasts, and protozoa, showed that they all have many reproductive phenomena in common, genetics became a field of research in its own right. Genes were of interest not only because of their intrinsic biological significance but also because of potential commercial and agricultural applications; industrial fermentations could be improved with engineered strains of yeast, and crops could be made better by genetic manipulations.[3]

From a biological point of view, the most obvious thing phages did was reproduce, fast and furiously. Thus, Muller's suggestion that phages might be simply naked genes came to fruition in the focus on phage reproduction as an ideal way to study this central biological problem.[4] As would soon become clear, not only did phages reproduce, but they did so with high fidelity. The remarkable stability of the gene, another key puzzle, was exhibited in phage reproduction as well. Clearly, viruses, including bacteriophages, had to be chemically and physically simple, perhaps only a single protein or aggregate of proteins (or, as some suggested, a nucleoprotein). Yet it was not so obvious in the decades immediately before World War II that microbes of any sort, whether bacteria, molds, or viruses, were going to be "the right organism for the job" of the biologist. Their small sizes, even with the best light microscopy, made conventional biology—traditionally dependent on morphologic and developmental information—difficult. Indirect studies, represented by formal genetics and chemical analyses, were not yet fully embraced by many biologists and were trusted by even fewer to provide clear understanding. Molecular structures, especially of large molecules such as proteins, carbohydrates, and nucleic acids, were vague at best. Genes were studied by mating experiments and thus thought to be somehow irrelevant to microbes without such sexual systems. It is worthwhile, then, to inquire why phages were of interest, even to the small group of pioneers in the late 1930s and early 1940s.

The history of phage research, right from the start, showed that scientists who identified themselves as "phage workers" struggled with diverse ideas and approaches that were only partially resolved by the time they were overtaken by the emergence of the more general disciplinary concept of molecular biology. Taking "physics versus biochemistry" has been used as a dividing principle, but this binary oversimplifies the diversity and complexity of both fields. Further, because the early phage researchers were few in number and located in only a few places, with spotty financial support, the law of small numbers likely explains the diversity of the original contributors to phage research. Thus, one group, at one university, might grow out of a biochemist's interest in phage, another might develop around a chemist, and a third might coalesce around a medical microbiologist. How these first tiny groups and, in some cases, isolated individuals came together to form a powerful and coherent research tradition is the theme we will now take up.

The Formative Groups

Although the earliest phage research in the third and fourth decades of the twentieth century was, for the most part, devoted to the clinical application of phage therapy, a very few scientists looked at phage as a biological puzzle. Surprisingly, for most, however, this interest did not stem from phage as models of viruses, since the status of phage as viruses (that happened to infect bacteria) was not of much concern to most biologists. Viruses such as tobacco mosaic virus, polio, and influenza were important because they were plant and animal pathogens, yet phage rarely entered the same discussions. Since the debate on the status of bacteria as models for so-called higher cells was not yet in full swing, it is likely that the viruses of bacteria suffered from the doubt that bacteria were similar enough to nucleated cells (eucaryotes) to yield useful comparative results.[5]

One investigator who did see phage as a model for animal viruses was Emory Ellis (1906–2003), a research chemist

at Caltech. In 1934, after receiving his Ph.D. in physical chemistry at Caltech, Ellis took a job with the Food and Drug Ad-
ministration, but, driven by the exigencies of the Great Depression in 1935, he took a job as an assistant to Seeley G. Mudd
(1895–1968), a physician at Caltech studying the use of Caltech's
new supra-voltage x-ray facility for cancer therapy.[6] Ellis was
hired to conduct basic research on cancer, and he decided
to study carcinogenesis in animals. At the height of the Great
Depression, however, animal experimentation—whether with
Rous sarcoma virus in chickens, rabbit papilloma virus in
rabbits, or even chemical carcinogenesis studies in mice—was
prohibitively expensive. Ellis hit upon phage as a model system to study "how viruses worked." Cheap, quick, and techni-
cally simple, phages provided Ellis with an ideal experimental
system to study basic virus biology. He clearly had no qualms
about equating the biology of phages with the biology of animal viruses. Ellis based his experiments on the original work of
Félix d'Herelle, whose research he studied carefully. Initially,
he set out to replicate d'Herelle's basic experiments to become
familiar with phage techniques and to verify for himself some
of d'Herelle's claims that were still controversial. He followed
d'Herelle's procedures and isolated some phages from Pasadena
sewage. He asked a fellow researcher for a bacterial strain that
was easy to grow, and when that colleague, Carl C. Lindegren
(1896–1987), provided him with *Escherichia coli,* Ellis selected
E. coli as his host bacterium. Thus, this first Caltech phage research set the direction toward the prototype organism around
which almost all early molecular biology would coalesce.[7]

Ellis's work on the basic way phages grew and lysed cells
interested Mudd. At the time, it appeared that various bacterial products (called "Coley's Toxin" after its discoverer) were
useful in arresting the growth of cancers in experimental animals.[8] Mudd thought that the phage lysates might provide a

better way of preparing effective toxins for treating inoperable cancers. Ellis was forced into two concurrent research pathways, his own inclination to find out more about the basic biology of phage and Mudd's program of treating mice carrying transplanted tumors with phage lysates. Ellis's career in phage biology was shaped once again by basic economic necessity pitted against his choice of scientific inquiry.

"Two Enemy Aliens and a Social Misfit"

Max Delbrück, one of the key founders of the APG, famously characterized the three young scientists who met in 1943 during the height of World War II to talk about phage as "two enemy aliens" (Salvador Luria, an Italian Jewish refugee, and himself, a patrician German visiting scholar) and a "social misfit" (Alfred Hershey, who liked living on boats, which seemed strange to Delbrück).[9] This meeting in St. Louis, Missouri, where Hershey was working in Washington University's department of microbiology, has been characterized as the origin of the APG.

Max Delbrück, by this point (1943) a bona fide immigrant from Germany, came from a distinguished German academic family, which included his father, Hans Delbrück, a famous military historian and politician, and his two brothers-in-law, Dietrich Bonhoeffer, an influential theologian, and Klaus Bonhoeffer, a jurist, both anti-Nazi martyrs. In his youth, Max was interested in astrophysics but eventually took his doctorate in theoretical physics under the direction of Max Born in Göttingen in 1930. After his degree, Max accepted an invitation to join the underappreciated physicist John Lennard-Jones (1894–1954) at Bristol, England, as a post-doctoral fellow. This was an opportunity to learn English well and expand his fields of interest to the chemical bonding physics of which Lennard-Jones was a pioneer. After a year in Bristol, Delbrück obtained

Figure 2. Max Delbrück *(left)*, with Salvador Luria at Cold Spring Harbor, about 1952 (Courtesy of the Caltech Archives)

a Rockefeller fellowship to return to theoretical physics, spending half a year in Copenhagen with Niels Bohr and half a year with Wolfgang Pauli (1900–1958) in Zurich. This international experience exposed the impressionable Delbrück to many of the exciting young (and not so young) physicists of the day.

Figure 3. Martha Chase and Alfred Hershey at Cold Spring
Harbor Laboratory, 1953 (Courtesy of Cold Spring Harbor
Laboratory Archives, New York)

Bohr, a physicist with a philosopher's heart, had a major influence on Delbrück. In contrast to the rather dour Max Born, Bohr was an expansive and outgoing questioner of all sorts of things. He was deeply interested in the epistemology of the newly discovered quantum world and his expansive view of how physics might be employed to gain new insights into other fields like biology seems to have aroused Delbrück's first interests in biology. For several years, Delbrück returned to Copenhagen at every opportunity, becoming a lifelong captive in Bohr's orbit. In 1932, Delbrück accepted a position in Berlin as the resident theoretical physicist in a group led by Lisa Meitner (1878–1968), partly because of its proximity to the Kaiser-Wilhelm Institute of Biology, where he was "entertaining friendly relations." Although he worked on physics with Meitner, Max also started to think about biology, organizing a regular discussion group in his childhood home in Berlin that examined connections between physics and biology, discussions strongly influenced by Bohr's views.[10]

The revolutionary feature of the "new physics" (that is, quantum action) of Planck was that energy is not continuously distributed: at the most basic level, a particle cannot have just any amount of energy but is restricted to discrete amounts of energy (a quantum of energy)—just as a cat can have one, two, or three kittens but not one and two-thirds kittens. Energies possessed by electrons came in such chunks, but by some mysterious process an electron could acquire more energy and then become stable again with more chunks of energy, so that it would have a different energy level. These energy transitions, observed in electrons in atoms, were rare, abrupt, and quite stable. It is no surprise, then, that the newly studied process of gene mutation might be viewed by a physicist as analogous to the transitions of energy quanta. Genes mutate (very rarely) from one discrete state to a different, but very stable, state with no intermediate states being possible. For Delbrück and his biophysics study group in Berlin, gene stability and mutation could be framed in a way that appealed to physicists trained in quantum mechanics and that suggested ways to study this biological puzzle. Biologists were brought into the discussion, and one product was a rather strange collaborative publication authored by a geneticist (Timoféef-Ressovsky), a radiation biophysicist (Zimmer), and a physicist (Delbrück) with the title (translated into English) "On the Nature of Gene Mutation and Gene Structure." This paper would become the locus classicus for the founding of molecular genetics in the canonical account of molecular biology. It made Delbrück one of the foremost gene theorists of his day, at least among physicists.[11]

As a young German academic, Delbrück's career path included the certification process to become a university lecturer, termed the "habilitation." In the mid-1930s, however, one's success at this certification was subject to political considerations as well as academic prowess; even though Max attended a program to instill "political reliability," he was not granted the habilitation, even on his second attempt in 1935. His disdain for the Nazis was too apparent. With an academic career in Germany in serious doubt, a new opportunity arose when he received a Rockefeller fellowship to go to Caltech in 1937.

Delbrück, a non-resident alien in the United States, arrived in America intent on extending his study of genes in the fruit fly with T. H. Morgan at Caltech. He found *Drosophila* genetics full of lore, jargon, and esoteric terminology, and he was not convinced that the Caltech approach to the gene through *Drosophila* chromosome linkage studies was going to be successful. In early 1938, when he missed a seminar by Emory Ellis on the phage work, he sought out Ellis for a private tutorial. Ellis later recalled this meeting:

> Did Max know about bacteriophage prior to meeting with me? I don't think so. Max came to Caltech (on a Rockefeller Foundation fellowship) primarily (I think) because of his interest in genetics and the unique properties of living matter, and because of the presence of Thomas Hunt Morgan at Caltech. . . . When Max came into my small lab room, I had just plotted the results of a successful phage growth experiment which showed two steps in the growth curve. I explained how the phage particles could be counted by petri dish assay, and how they multi-

plied in the bacterium, then were released into the culture by destruction of the bacterial cell. At first he was skeptical but returned and asked to join in the work. This joint effort led to the seminal paper "The Growth of Bacteriophage."[12]

As Delbrück recounted their meeting,

> I was absolutely overwhelmed that there were such simple procedures with which you could visualize virus particles. I mean, you could put them on a plate with a lawn of bacteria, and the next morning every virus particle would have eaten a macroscopic one-millimeter hole in the lawn. You could hold the plate up and count the plaques. This seemed to me just beyond my wildest dreams of doing simple experiments on something like atoms in biology, and I asked him whether I could join him in his work, and he was very kind and indeed invited me to do so.[13]

After another year of phage work, in the fall of 1939, with the German invasion of Poland and the start of World War II, Delbrück wanted to remain in the United States; without further Rockefeller support and with no position available at Caltech, Delbrück appealed to Warren Weaver, director of the Rockefeller Foundation, for help. After some uncertainty, Max found a position teaching introductory physics at Vanderbilt University in Nashville, Tennessee, a position that still allowed him to continue his phage work in collaboration with colleagues in the School of Medicine.

By the time Delbrück and Ellis had completed their first phase of the phage work that convinced Delbrück that phage would be an ideal tool to study fundamental problems of reproduction and genetics, Delbrück had departed for Vanderbilt and Ellis was pulled back to Mudd's tumor work. That work, however, would soon end with the urgency of war research at Caltech, which provided Ellis an escape from an increasingly uncomfortable collaboration with Mudd. In one of those contingencies of history, Ellis abandoned phage research to put his physical chemistry background to use as one of the leaders who established the Naval Ordinance Test Station at China Lake, where he contributed significantly to the U.S. guided missile program.[14]

Meanwhile, in Paris, young Salvatore Luria (upon immigrating to the United States, he changed his name to Salvador) was looking for a way to get to America. Luria, an Italian from Turin, had trained as a physician in Italy, but he was drawn to basic laboratory research and spent a period in Rome combining training in radiology and laboratory work on radiobiology of phages with the Italian phage worker Geo Rita. As a Jewish refugee, Luria made his way to Paris where he was able to continue his radiobiological studies in the Radium Institute, publishing on the sizes of phages determined by target theory approaches. Having read the first paper on phage by Ellis and Delbrück, Luria wrote to Ellis, assuming that he was the head of the laboratory, inquiring about a position with him at Caltech. Ellis, rather insecure himself, wrote back that he could no longer work on phage and, indeed, would soon be looking for a position himself. Luria arrived in America in September 1940, unemployed. However, a friend from his youth in Italy, Enrico Fermi of later atom bomb fame, then working at Columbia University, recommended him to Leslie C. Dunn (1893–1974), an influential geneticist who had read Luria's radiobiological work with Holweck. Dunn secured a position for Luria at Columbia where he collaborated with Frank M. Exner (1899–

1990), a physicist who had just built a new megavoltage x-ray machine.[15]

Although Luria had no luck with Ellis, he managed to correspond with Delbrück and the two arranged to meet in late December 1940 at the annual meeting of the American Physical Society in Philadelphia. Although at this point the United States had not entered World War II, both Luria and Delbrück, as "enemy aliens," were subject to the newly enacted Alien Registration Act of 28 June 1940. This act, usually known as the "Smith Act," formed the basis for President Roosevelt's infamous executive order EO 9066 of 16 February 1942 that allowed relocation, internment, or other restrictions on individuals in the United States with varying ties to Italy, Germany, or Japan. In the end, at least 10,000 people with Italian heritage were affected, while a vastly larger number of people of Japanese heritage were forcibly interned and deprived of their homes, livelihoods, and civil rights in the name of national security. Although Delbrück was in the United States as a German citizen who could return to Nazi Germany, being "non-repatriated and non-Jewish," he had decided to remain in the United States, if at all possible. Luria, on the other hand, was a Jewish refugee and was, to a greater extent, subject to the United States immigration policies. Although Luria later seemed to minimize his concerns about his status as an enemy alien, many in the Italian community were harassed and some imprisoned or forcibly "relocated" to regions of the United States away from the East Coast, where it was feared they might be in a better position to undertake pro-Axis sabotage.[16]

That both Delbrück and Luria relocated from their transient posts on the West and East Coasts, respectively, to faculty positions in top research universities not far apart in the midwest may have been simply fortuitous, but there are indirect indications of significant behind-the-scenes machinations on the part of their supporters at the Rockefeller Foundation to allow them to continue their phage research program, as well as to position them out of the way of the growing anti-alien hysteria of wartime America. At the expiration of Delbrück's fellowship at Caltech in September 1939, Morgan, the chair of the biology department, wrote to the Rockefeller Foundation pleading for help in finding a suitable position for Delbrück since Caltech had no available funds to keep him on. Within a couple of months, Max had received an offer of an instructorship in physics at Vanderbilt University. Vanderbilt had been looking for a "German physicist" to bring the new "German" quantum physics into their curriculum, and in exchange they agreed that Delbrück could devote his research program to phage. This physics instructorship was initially funded for several years by the Rockefeller Foundation. The relationship between the Rockefeller Foundation and Vanderbilt was a particularly close one, so it is not surprising that such a mutually advantageous arrangement was rapidly concluded. Max arrived in Nashville in January 1940.[17]

Luria, meanwhile, was also supported by the Rockefeller Foundation, starting with his arrival in the United States in 1940, and this support required him to accept the first "respectable" academic position he was offered. This offer came in September 1942 in the form of an instructorship at Indiana University in Bloomington, Indiana, a place totally unfamiliar to Luria at the time. In January 1943 he arrived in Bloomington, only a long day's drive from Delbrück in Nashville.[18]

The third member of what might be called the nucleus of the APG was Alfred Hershey, a protégé of Jacques Bronfenbrenner, one of the few early phage workers interested in the basic nature of phage rather than its therapeutic potential. Bronfenbrenner was a distinguished bacteriologist, a disciple of the Rockefeller Institute microbiologists Hideyo Noguchi

(1876–1928) and Simon Flexner; he became an early adherent of d'Herelle's particulate theory of phage in about 1925. After becoming an associate member of the Rockefeller Institute, Bronfenbrenner took up the chairmanship of the department of bacteriology and immunology at Washington University in St. Louis in 1928. In 1934, he recruited Alfred Hershey, fresh with his Ph.D. in chemistry and bacteriology from Michigan State University, to join Washington University as an instructor in his department. Until about 1940, Hershey worked on bacterial growth phenomena (an extension of his doctoral work), but he then joined Bronfenbrenner in studies of the nature of phage using quantitative immunological approaches.

Hershey's careful mathematical and quantitative studies, and perhaps the relative proximity of St. Louis to Nashville, impressed Delbrück, who wrote to Hershey in late 1942 and invited him to visit Nashville and give a talk about his phage work to Delbrück's small, informal, but interdisciplinary discussion group at Vanderbilt. The visit was an apparent success, perhaps an attraction of opposites. Delbrück famously described Hershey in a letter to Luria in February 1943: "Drinks whiskey but not tea. Simple and to the point. Likes living in a sailboat for three months. Likes independence." When Hershey asked how to prepare for his talk, Delbrück, in his most Olympian manner, again provided an epigrammatic answer: "The speaker should assume complete ignorance and infinite intelligence on the part of the audience."[19]

Nucleation: The First Phage Meeting

Hershey reciprocated with an invitation for Delbrück to visit him in St. Louis. In April 1943, both Luria and Delbrück joined Hershey there for what would be remembered as "the First Phage Meeting," later described by Max as a meeting of "two

enemy aliens and one social misfit."[20] These three would ar-
range further visits, and in Luria's words, "[Hershey] immediately became a partner in our phage adventure. . . . Al formed
with Max and me the nucleus of the 'Phage Group,' a nucleus
that grew slowly at first, then at an almost catastrophic rate as
more people realized the remarkable opportunities of bacteriophage as a research object." Always one to look ahead, Delbrück urged Hershey to join him for a while in Nashville in
the fall of 1943 so they might do a few experiments together:
"My guess is that there will be a great rush on phage work after
the war, so we better settle the elementary problems now, *that
we may speak with authority later*" [emphasis added]. There
can be no better indication that Delbrück envisioned a future
community of phage workers that must be molded and guided
into the proper channels. In short order he would set about
developing, codifying, and policing the framework of basic
commitments that such a phage community should come to
share. As Hershey later recalled, "he devoted great effort and
intelligence to encouraging, appreciating, and steering the work
of others, probably often at the expense of his own. That generosity of spirit was one of his chief characteristics."[21]

Perhaps the timing was just good fortune, but in any
event, a lucky break came in 1943 when Luria devised an experiment that he thought might illuminate something very basic
about the way mutant organisms arise. Delbrück immediately
saw the elegance of Luria's idea and developed the mathematical approaches to interpret the experiment of the method later
known far and wide among geneticists as the "Luria-Delbrück
Fluctuation Test." This result was published in *Genetics,* the
widely respected journal of the Genetics Society of America.[22]
Others had attempted to test whether mutations were *caused*
by the selective agent used to demonstrate their existence, or,
conversely, were pre-existing from some unknown source and

only *selected* by the experimenter. Luria and Delbrück managed to use the phage growth experiments to answer this important question definitively on the side of the selectionists. This result was central to Darwin's theory of natural selection, and hence it immediately made phage biology something worth noticing.

Luria and Delbrück followed up on d'Herelle's earlier work on the fact that phage-resistant bacteria appear in a culture of bacteria a while after the culture is originally lysed by infection with phage.[23] These bacteria appeared to be mutants of the original bacteria and thus might be useful tools to study the process of mutation of bacteria itself. The classic problem confronted them: did the phage cause the mutation or merely select against the non-mutant forms so the phage-resistant bacteria could outgrow them? As Luria tells the story in his autobiography, he got an idea for a new experimental approach while watching someone play a slot machine at a faculty party at Indiana University. The slot machine was in a way like the bacterial culture: usually it paid no return, but rarely paid a jackpot. Luria realized that if one had many small samples of bacteria, most samples would have no phage-resistant mutants, but a few might have jackpot mutations, and the size of the jackpot would vary with how many bacterial cell divisions occurred from the time of the mutation.

He envisioned a simple experiment of growing many small cultures starting from one (or a few) bacteria; then after several generations testing how many phage-resistant bacteria (if any) were present, one could (with some simple statistical calculations) determine the rate at which mutations were occurring in the *absence* of the selecting agent (phage). The statistical model, based on the Poisson distribution, predicted a wide fluctuation in the number of mutations in the various cultures, if the mutations were occurring at random in the absence of selection. If, however, the mutations were being caused by the selection (phage killing), the cultures would be very similar with very little fluctuation in number of mutants per culture. The experiment worked spectacularly well, in favor of random mutation in the absence of selection (Darwin over Lamarck). This experiment, frequently called the "Luria-Delbrück fluctuation test," embodies two key features of the new approach to biology taken by the APG: first was the simplicity of the experiment, the treatment of the biological system as a simple black-box, input-output system (the slot machine metaphor is telling); and second, the crucial role of statistical theorizing as central to the interpretation. Interestingly, the collaboration reflected another aspect frequently found in the APG: the collaborators worked in a joint way that did not track neatly by individual expertise.

Luria recounts his intellectual pathway to this key experiment in his autobiography, where he explains that he was taking on the theories of Sir Cyril Hinshelwood (1897–1967), a Nobel Prize–winning chemical kineticist, who essentially denied the existence of genes in bacteria. Hinshelwood published heavily mathematical papers claiming that what were being called mutations were simply different quasi-stable kinetic states of bacterial metabolism. While the Luria-Delbrück experiment did not directly falsify Hinshelwood's kinetic theory, it challenged his views in significant aspects, and Hinshelwood's ideas are now in the dustbin of history. Soon, two other microbial geneticists, Esther Zimmer Lederberg (1922–2006) and Joshua Lederberg (1925–2008), devised the "replica plating technique" that would also support the selectionist camp (and Darwin as well).[24]

As the tiny phage nucleus pushed forward in the early 1940s with Delbrück's program of "settling the elementary problems," it became apparent that different phage isolates behaved somewhat differently: plaque size, host range, plaque appearance, time to lysis, burst size, and so on. Some agreement on uniformity was needed so that comparisons between laboratories would be possible. In the summer of 1944 at the Cold Spring Harbor Laboratory, Delbrück urged the small group of phage workers that had started to coalesce around him, Luria, and Hershey to agree on a standard collection of phages and host bacterial strains that they would all share and use exclusively. This agreement became known in phage lore as the Phage Treaty of 1944. The host strain that was adopted (*Escherichia coli* strain B) was that used in the Luria-Delbrück fluctuation test work of 1943, and it has been traced back to one of d'Herelle's strains provided to Jules Bordet in 1920, which made its way to Luria, probably through Hershey from Bronfenbrenner, who got it from Bordet in 1924.

Seven phages were adopted as standards from among the collection of phages familiar to the small group of phage workers. All these phages were lytic on *E. coli* B. One feature that seemed to be important was that, for each phage isolate, a mutant form of the host, *E. coli* B, could be found that was resistant to infection (that is, no phage plaques would appear) by each individual phage. This phage-host system of sensitivity and resistance provided a simple identification of each phage, something that was crucial for phage research at this early stage. Some of the phages were already known; for example, phage T1 was identical to phage α and phage P28 used by Delbrück and Luria, and by Bronfenbrenner and Hershey, respectively. Phage T2 was identical to phage γ and phage PC used by these same workers. Other phages were freshly isolated from samples of phages being used for phage therapy: phages T3, T4, T5, and T6 were isolated from a mixture provided by Tony Rakieten (1902–1984), a former collaborator of d'Herelle from his time at the Yale School of Medicine in the early 1930s; and phage T7 was from a therapeutic phage preparation from

Ward J. MacNeal (1881–1946), a physician at the New York Post-Graduate Medical School and Hospital. This set of seven phages and their single bacterial host strain came to be known as "Snow White and the Seven Dwarfs" of the APG.

The Phage Treaty of 1944 represented an early step in the development (and enforcement) of a key shared commitment that would define the nascent APG. This commitment to the so-called T phages (T for type) would be challenged almost immediately by the phenomenon of lysogeny and phages, notably lambda bacteriophage, that did not always lyse the host cell. By the 1960s, other phages were identified that provided interesting properties for study: single-strand circular DNA of phages $\varphi X174$ and $S13$; phages with RNA genomes such as $R17$, $Q\beta$, $MS2$, $f2$, and others; phages that attach to specific bacterial surface structures, for example, phages $M13$ and fd; phages that cause cell genes to rearrange, such as phage mu; and phages such as $P1$ that frequently carry host genes between bacteria in an important phenomenon called "horizontal gene transfer." By this time, however, the group's authority had become fragmented, as the APG merged into the broader field of molecular biology and the value of other phages was recognized for their specific, useful, properties.[25]

Standardization of experimental material was just one way this tiny nucleus of phage workers started to build consensus among the initially isolated phage community. As discussed in some detail later, another way was through educating and indoctrinating recruits who participated in a training course first started at the Cold Spring Harbor Lab in the summer of 1945. Initially, Hershey, Luria, and Delbrück focused their attention on their immediate colleagues and students to develop a community of like-minded phage workers. During the war years, universities and academic research were disrupted in many ways: research programs were redirected toward war-related matters; eligible young male students were conscripted; funds from philanthropic sources were in short supply due to both the war and the lingering Great Depression; and transatlantic scientific communications and other international collaborations were severely compromised. Immediately after the war, however, belief in "science" blossomed, students returned to campuses, European visitors started to appear in North America, and, importantly for biological research, new technologies such as radioisotopes and chromatographic separations appeared as wartime byproducts.[26]

It is important to note, too, that in addition to the small group surrounding Hershey, Luria, and Delbrück there were several other scientists in North America who, although rather isolated from each other, both geographically and intellectually, had taken up phage research for their own particular reasons. These groups represented targets of recruitment for Delbrück in his drive to put phage on the map. This effort was hindered, however, by existing disciplinary boundaries, barriers that would plague the APG for its entire existence, but which would eventually succumb to the dominant paradigm of molecular biology. These phage workers had diverse interests in phage, some more closely allied with the nascent APG than others.

Probably most distant were microbiologists who recognized in the 1930s that the host specificity of phages could be exploited as a new method for characterizing different strains and species of bacteria.[27] So-called phage typing researchers viewed phages instrumentally, often with little regard for the basic biology of phage. They were immersed in the diversity of phages, not concentrating on a small set of phages that infected a single strain of *E. coli*. Another group of phage enthusiasts were a collection of radiobiologists intrigued by the ease with which phage allowed the study of basic mechanisms by which radiation damage affected biological material. The atomic physicists at Yale, led by Ernest Pollard, exemplified this group. Their approach to phage research, of course, was not much different than that of the founders of the APG, who initially saw radiation target theory as a key tool in their research agenda. They saw phage as a way to study radiation effects rather than using radiation as a way to study phage biology. It was mainly a matter of primary versus secondary focus, and thus these phage workers became closely allied with, but not incorporated into, the APG quite early. A third strain of phage research that would eventually be incorporated into the APG as well was represented by Macfarlane Burnet in Australia and the senior Wollmans in Paris, who were keenly interested in lysogeny as a problem in host-virus interaction, a problem that was initially eschewed by the APG as an unnecessary complication. Later, temperate phages and lysogeny would be rather grudgingly accepted as an important part of the APG agenda. The most problematic phage work was being done by scientists who identified as chemists and biochemists. The two prominent leaders of the biochemical study of phage were John H. Northrop and Earl A. Evans, Jr. They agreed on the main goals of the APG but disagreed on the fundamental experimental approaches and, in Northrop's case, on the basic nature of phage itself. These biochemists were able to interact with the APG, but they usually did not feel themselves to be part of the group. Available evidence suggests that the feeling was mutual.

John Northrop of the Rockefeller Institute was a protein chemist and winner of the 1946 Nobel Prize in Chemistry for his work on crystalline enzymes. Northrop had discovered the auto-catalytic nature of some enzymes such as pepsin (which is formed by action of its precursor, pepsinogen, on pepsinogen itself). He viewed the growth of bacteriophage as another case of auto-activation (rather than de novo reproduction) and had obtained a small phage in crystalline form in 1938. His 1939 book *Crystalline Enzymes: The Chemistry of Pepsin, Trypsin, and Bacteriophage* clearly indicated that he classified phage as an enzyme rather than as a microbe. His protégé Alfred P. Krueger (1902–1982) at the University of California, Berkeley, was a staunch defender of his mentor's theory of phage; he debated Ellis and Delbrück on several occasions. Neither Northrop nor Krueger seemed to have any desire to associate with the nascent APG, and they steadfastly rejected the basic framework of shared research commitments being developed by the phage workers around Delbrück, Luria, Hershey, and their ilk. Another of Northrop's younger protégés, however, Seymour Cohen, fully engaged with the APG, although he would never admit to being a "member" of the group. The other group that took a biochemical approach to phage research was that of Earl Evans, Jr., at the University of Chicago. Evans was chair of the department of biochemistry and a well-known contributor to studies of intermediary metabolism, an early proponent of the use of radioisotopic tracers in metabolic studies immediately after World War II.[28]

Both Cohen and Evans saw in the striking reproduction rate of phage and the ease of their assay a new approach to the vexing problem of protein synthesis. In the 1940s, it was a complete mystery how amino acids were polymerized into fully formed proteins of specific but vastly diverse composition (specificity of sequence would await Frederick Sanger's [1918–2013] sequencing of insulin in 1949). With the availability of isotopically labeled precursors of proteins and nucleic acids, optimism was high that study of the flow of label from medium to cell to phage progeny would illuminate the pathways by which these biopolymers were synthesized. The chemical nature of the gene and its faithful duplication during every cell division was still not clear (in spite of the work done by Avery, McCarty, and McLeod on bacterial transformation).

Evans was an early participant in some of the APG meetings, and he took the phage course in 1946 (along with other biochemists Cohen and Mark Adams). Later his group was represented by two younger colleagues, Lloyd Kozloff and Frank Putnam, who along with Cohen would be a regular biochemical presence at APG events. Indeed, it was probably these three who eventually overcame some of Delbrück's initial prejudice against biochemistry by their important contributions to fundamental phage research in work that supported the early shared commitments of the APG.

Although these diverse and scattered groups of phage researchers contributed to the APG by accretion, the APG's main growth came from the physics community, partly from Delbrück's proselytizing and partly from the conditions in the postwar physics community. In the immediate aftermath of the war, it was clear that much of physics would be dominated by atomic and nuclear physics, with expensive equipment, big research groups, and government involvement. Indeed, the development of the National Laboratories at Oak Ridge, Brookhaven, Argonne, and Los Alamos presaged this. In addition, there was unease in the academic community at large about nuclear physics research in the aftermath of the bombings of Hiroshima and Nagasaki.[29] Many physicists were looking for other areas of research. Indeed, the government recognized the need to address this broad public concern when it initiated the "Atoms for Peace" propaganda program and the "Plowshare Program" proposing the use of nuclear explosions for peacetime construction projects. Physicists could, with very little expensive equipment, use their theoretical training and understanding of the basic physics of radiation biology to carry out fundamental biological studies on phage, which was being advertised by Delbrück as the "atoms of biology." In contrast to much of traditional biology, such as fruit fly genetics, phage research was not burdened with masses of historical results to learn, complicated manipulations, or new mathematics. As has often been noted, phage became the physicists' "black box" or "gadget."

One such physicist was Cyrus Levinthal (1922–1990), who had just finished his dissertation at the Radiation Laboratory at Berkeley in 1950 and accepted a position in the physics department at the University of Michigan. Levinthal went from the home of the cyclotron to a center of research using the new electron microscope, which was being developed by Ralph W. G. Wyckoff (1897–1994) and Robley C. Williams (1908–1995). Wyckoff and Williams had connections in both the department of physics and the Public Health Virus Laboratory, and they pioneered the metal evaporation shadowing technique that they applied to study viruses. A combination of three factors perhaps led Levinthal to turn his attention to phage research: the availability of this relatively rare electron microscope technology in his department, his experience in physics related to the target theory approaches to biology, and the proximity of Luria, a target theory phage biologist, by this time in Urbana, Illinois. Levinthal's first paper on phage used extensive electron microscopic imagery, and he thanked Hershey, Luria, and Luria's Illinois colleague Martha Barnes Baylor (1915–2006) for help with phage biology. He sharpened his phage laboratory skills at the phage course at the Cold Spring Harbor Laboratory in the summer of 1952. Levinthal went on to become a fully vetted member of the APG: a doctorate in modern physics, a graduate of the phage course, and a willing acolyte to the initial APG triumvirate.[30]

At the other end of the career spectrum, one might consider the example of Leo Szilard, already a distinguished physicist prior to full-time commitment to (proto-) molecular biology. Like some others, he was an iconoclast and developed a

somewhat fraught relationship with the APG while still maintaining close connections. Szilard, the inventor of the nuclear
chain reaction, joined Fermi on the Manhattan Project at the
University of Chicago at the "Metallurgical Laboratory" (a label
chosen to disguise its real work on the atomic bomb), where
he collaborated with the young physical-organic chemist Aaron
Novick. After the war, Szilard and Novick, motivated both by
scientific interest as well as moral and philosophical rejection
of the future of nuclear research in the United States, joined a
new unit at Chicago, the Institute of Radiobiology and Biophysics. Here they started a program of research on basic biology of growth using microorganisms. As a center for much of
the Manhattan Project, Chicago at the end of the war was home
to several notable physicists and chemists besides Szilard. James
Franck and Harold Urey (1893–1981) were also turning their
eyes toward biological problems.[31]

One gets the impression of a "phage zeitgeist" in the middle of the 1940s, as multiple factors emerged that facilitated
such a nascent endeavor: brash young physical scientists looking to put war research behind them; stagnation of classical
biology; new technologies such as isotopic tracers, ultracentrifuges, and electron microscopes, which made new experiments
possible; top-down support and encouragement from patrons
such as the Rockefeller Foundation as well as the federal government; and the appearance of key proponents with a particular combination of talents for discipline building.

By the late 1940s, the phage course had gained a foothold
in the summer schedule at Cold Spring Harbor and was becoming a crucial site of recruitment and indoctrination; periodic meetings and newsletters defined a community of scientists who saw themselves as part of a growing and legitimate
network; and age-old biological questions were being answered

by the simple yet elegant experiments based on the new microbial experimental materials.

But how was this group organized? How did one join? How can we identify "membership"? As we will examine in detail next, participation in the phage course was certainly a major plus to be accepted into the APG, but it was neither sufficient nor necessary. Delbrück would screen applicants, famously by an "entrance" exam based on simple mathematics of using exponential scientific notation to work with large numbers, but he also had benchmarks about student backgrounds, their mentors, and their allegiance to phage. As Jean Weigle, one of Delbrück's colleagues, noted: "I knew that, to use Max Delbrück's disdainful indictment of phage work which did not come up to his standards, the Lederbergs, the specialists par excellence of bacterial crosses, 'had never taken the phage course.'"[32]

Delbrück curated the mailing list of his newsletter, *Phage Information Service,* to add and delete individuals over time, apparently deciding who would be "in the loop" and who would not. Certainly, too, correspondence between Luria and Delbrück includes derogatory remarks about the work of some scientists, who nevertheless continued to be invited to the phage meetings and were regular recipients of the newsletter. Three longtime phage researchers provided related perspectives on the criteria for membership in the APG. Seymour Cohen, the phage biochemist, reported that he was a regular participant in the phage meetings and other group activities, but he adamantly insisted he was not a member of the APG. The self-appointed historian of the APG, Gunther Stent, confirms the APG view of Cohen: "Cohen was, and yet was not, a member of the APG. He was in close contact with the work of the Group's members, used the experimental method they had developed, and even collaborated with some of them. Yet, he felt he was an outsider because Max and his disciples did not sufficiently appreciate the capital importance of his biochemical approach that Cohen himself attributed to it."[33]

Lloyd Kozloff, from the Chicago group of biochemists, commented that being a "member" meant that one sent one's manuscripts to Delbrück for approval and comment prior to submission for publication. Wacław Szybalski (1921–2020), a microbial geneticist with Demerec at Cold Spring Harbor at this period, distinguished the APG from the biochemists by the belief (on the part of the APG) that the biochemists published too many papers on minor points.[34] Doing important phage work while being a biochemist was not enough for membership in the APG. Fidelity to certain shared beliefs was necessary, but sometimes, as in the case of Seymour Cohen, it appears that *rejection* of some beliefs was also required.

Neither did being a "physicist doing biology" automatically make one a candidate for the APG. Some of those renegade physicists described earlier, despite having objectives and philosophical inclinations that overlapped with the APG, never fully embraced the framework of shared commitments of the APG. Some dropped in or dropped out as the occasion required, as Szilard did, and others, like James Franck, went their own ways entirely.

For the most part, microbiologists seemed more comfortable with their evolution into phage biologists. The basic approaches they had developed, such as dilution, colony counting, target theory methods, and growth curves were familiar, as was the concept of the organism as an individual, (nearly) invisible thing with rudimentary chemical characterization. Stuart Mudd (1893–1975) and Hershey found their prior studies of bacteria quite compatible with their new investigations in phage. This was not universally the case, however. The discoveries of George Beadle (1903–1989) and Edward Tatum (1909–1975) with *Neurospora* and the linkage of metabolic biochemistry with genetics in the early 1940s with their "one gene, one enzyme hypothesis" were derided by Delbrück as not germane to the fundamental question of the APG:

> Also the so-called biochemical genetics, the *Neurospora* genetics, that tied together genetics and biochemistry so beautifully, only highlighted the difficulty even more. You could learn an enormous amount about actual biosynthetic chains and their interrelations, but you did not learn at all about how the enzymes came about; and if you say "One gene, one enzyme," then the question remained, how does the gene make the enzyme, and how does the gene make the gene, and this was not in fact answered at all by any of the biochemical approaches. So in a sense I think my reservations about biochemistry were appropriate.[35]

Looking at this tiny group of phage enthusiasts from the early 1940s, we can see certain commonalities that drew them together, including their belief that more research workers were essential to their scientific progress and including their sense of responsibility to recruit like-minded workers to their community. It is tempting to identify Emory Ellis as the initial seed from which sprouted the APG. His vision of phage as a model organism, his approaches based on his background in physical chemistry, and his commitment to clear, quantitative experimentation all resonated with Delbrück and Luria, whose first American phage experience was through Ellis.

Ellis recounts how he and Delbrück met with Albert Krueger, Northrop's protégé, in futile attempts to persuade these established phage workers to come around to their framework of research commitments. Although circumstances precluded Ellis's continued phage research, Delbrück, Luria, and Hershey moved on from the early associations described in this chapter to organize, proselytize, and refine their common assumptions about phage research as the APG took shape in the next decade. Their role has been canonized by a later phage worker, Franklin Stahl (b. 1929), who wrote, "The Phage Church, as we were sometimes called . . . was led by the Trinity of Delbrück, Luria, and Hershey. Delbrück's status as founder and his *ex cathedra* manner made him the pope, of course, and Luria was the hard-working, socially sensitive priest-confessor. And Al [Hershey] was the saint."[36]

7

Building the Group

People, Place, and Paradigms

How does one build a scientific discipline? The development of the American Phage Group provides perspective on this important question. As we shall see, the APG evolved from the beginning with a focus on three key essentials: people, place, and paradigms. Although several individuals were involved in the development of what became the APG, it was Max Delbrück who was most conscious of the sociopolitical factors needed to nurture and advance the field. Starting in 1945 he initiated what became an annual training course at Cold Spring Harbor Laboratory (the Phage Course) with the express purpose of proselytizing and codifying what he saw as the central issues of phage biology. Through this course and the development of an informal network of communication, strengthened by the more formal *Phage Information Service* newsletters and an annual meeting of anointed acolytes, Delbrück and his small nucleus of colleagues effectively defined what it meant it to do phage research in the important decades right after the war. For Delbrück, it was vital to develop the specific elements of the phage research framework and to build consensus and commitment to these elements. In this way he could create a new formative group with a novel identity, not found in the existing disciplines of physics, chemistry, or biology.

He saw the need to focus on a particular set of problems using common techniques and uniform materials, so that results could be easily compared and materials exchanged freely between laboratories. Even more important, he defined a small set of problems that he believed were crucial to future advancement of the field. Naturally these problems and approaches reflected his own research background and pursuits. As we saw in his early correspondence with Hershey, he realized the importance of establishing the group's authority right away.

Of the small band of phage fans in the immediate postwar period, Max Delbrück was clearly the most entrepreneurial. Hershey was notably introverted, seen by some as socially awkward; he was still chafing under the heavy hand of Bronfenbrenner at Washington University and given to frequent migraines. Luria was settled at Indiana University, but, as a recent Jewish émigré scientist, he was struggling to develop his research program in a classically oriented midwestern biology department. In addition, Luria battled periods of near clinical depression.[1] Delbrück, on the other hand, had the advantage of an illustrious German academic pedigree and the confidence born of connections with such world-renowned scientists as Bohr, Morgan, and a host of prominent European physicists. He had been raised in a patrician family with a sense of noblesse oblige. He was the ideal person to undertake the de facto leadership of the new enterprise.

One cannot ignore the blind luck that sometimes connects people, ideas, and geography. The role of the Cold Spring Harbor Laboratory as an incubator for the newly hatched APG was central to its success and will be examined more exten-

Figure 4. Max Delbrück leading a phage discussion at Caltech, 1949. *Left to right:* Jean Weigle, Ole Maaløe, Elie Wollman, Gunther Stent, Delbrück, and Giorgio Soli. (Courtesy of the Caltech Archives)

sively later. It provided, through the vision of its director, Milislav Demerec, a home for both research collaborations and discipline building. Although there had been small, sporadic meetings of Delbrück, Hershey, Luria, and a few other people at several midwest locations, it was Cold Spring Harbor Laboratory that became, in essence, the official home of the APG in 1945. The availability of rentable lab space for summer researchers, following a long tradition of "research stations," allowed Delbrück, Luria, and a few students to work together for extended periods and to build consensus on key aspects of the framework that would become the APG. As Delbrück noted, it was most efficient to have a short training course in "standard" methods of phage research for new researchers, and thus was born the famous phage course that would play a key role in the origin of molecular biology.

Delbrück's own situation in the physics department at Vanderbilt contributed to some of the impetus for summer research at Cold Spring Harbor. Although Delbrück seemed to be well regarded at Vanderbilt, he was certainly an outlier doing biological work in the physics department, and he expressed a dim view of the pool of potential graduate students who might join him there. Summers at the Cold Spring Harbor Laboratory—with no teaching duties and colleagues who gathered from several places to make common cause in phage research—provided Delbrück with an ideal way to test out ideas and develop strategies for expanding phage research. Demerec was impressed enough to try to recruit Delbrück as a permanent biophysical addition to the Cold Spring Harbor Laboratory staff, to replace Ugo Fano who was eager to get back to theoretical physics and who was about to move into war work at the Aberdeen Proving Grounds. With Delbrück, Cold Spring Harbor Laboratory would get a physicist doing biology but also a dedicated experimentalist, something Fano was not. In correspondence with Demerec, Delbrück laid out his visions for the institution and, more expansively, for the field of phage research, a manifesto for the APG:

> In reply to your letter of October 23, enquiring whether I would be interested in a staff position at the Carnegie Institute at Cold Spring Harbor, I would like to submit at some length my views about the ways in which in my opinion a healthy program of research in this field might be organized. Phage

research falls into two branches: phage as a model for viruses in general, and phage as a tool for bacterial genetic research. Although these two directions seem to be divergent, they can, in fact not be pursued separately, for reasons which are familiar to anybody who is working in the field. Research in phage calls for the gradual training of highly qualified research workers who can introduce this research at other institutions. I am anxious to foster such development, because, unless the field of phage research is really opened up, the promising results obtained up until now will remain isolated curiosity pieces and will once more sink back into the oblivion of small print addenda in text books.

In a small way the Phage Course this summer was of some help in this direction, although perhaps only Martha Taylor [Baylor?] will actually continue to work on phage. . . . Such a course should continue to be part of the program, to permit people to get acquainted with phage research. However, the real training that is called for is the training that E. H. [Edward Herman] Anderson [1909–1952] and [August] Gus Doermann [1918–1991] have been getting. This kind of training takes about two years, but would vary from case to case.

. . .

To sum up, I consider it of equal importance to bring new men into phage research as to do the research ourselves. . . . more or less mature research workers like Luria, [T. F.] Anderson [1911–1991] and myself may stumble in the field by accidental contact at the moment when they are on the lookout for a new field. . . . or they may come to it at an im-

mature stage immediately after their Ph.D., before
they have established a line of their own. . . . or they
may come as graduate students for the Ph.D. . . .
I understand that Mrs. [Evelyn] Witkin [b. 1921] is
working at Carnegie under such an arrangement,
but I also understand that this was made possible by
her working simultaneously on the OSRD contract.[2]

Delbrück raised the question of a cooperative degree program
with Columbia; however, a formal program for graduate student research at the Cold Spring Harbor Laboratory would not
happen for several more decades. In the end, Delbrück held out
for something more stable than Demerec could offer.

The Phage Course

In the accumulated lore of phage research, nothing stands out
quite so strongly in the campfire stories of the old days as the
role of the phage course. Education, of course, is intellectual
seduction as well as social indoctrination, and scientific courses
have been used to inform but also to define the outlines of a
specific field. Such courses often were held at locations that
would become associated with a particular discipline or a specific scientific school. Frequently, they served as continuing
education for school teachers.[3] Occasionally, a course evolved
into a recognized entry into a developing discipline. Courses
in embryology at Woods Hole Marine Biology Laboratory became famous, as did the microbiology course at Pacific Grove
in California, presided over by C. B. van Neil.[4] Cold Spring
Harbor Laboratory offered summer courses as well, starting in
1927 with general physiology. These courses frequently would
be used to explore and evaluate elements of a particular research framework, and they effectively introduced and incul-

cated these elements in students who would become members of the formative group for a new discipline. The phage course was an example par excellence of this process.

As the geneticist Millard Susman (b. 1934) remembered, "the Phage Course was a rite of passage through which ordinary microbiologists could become members of the APG, a rather exclusive circle with Delbrück at its center."[5] Delbrück later demurred to Luria as the one who conceived of such a course:

> . . . why did we give this course? I think Luria was the promoter of that. Luria thought that if phage ever was to become an important line of research, and its potential really developed, more people would have to be brought into it. And therefore, one should make an effort to bring more people into this way, by giving the course. I think it was Luria more than I, but I may be wrong. I don't know. Anyhow it helped, even though only a few of the people who took the Phage Course actually became phage workers. At least this way we recruited quite a number of people who could read the phage literature with understanding.[6]

Delbrück, still at Vanderbilt in 1945, conducted a trial run of the phage lab work with two of his students in Nashville and then set off for Cold Spring Harbor Laboratory in the summer of 1945 for what would be the first of many iterations of the phage course, which would run until 1970.

Were cosmic forces at work during the first two weeks of August 1945? A German-American physicist, a student of the co-discoverers of uranium fission (Lisa Meitner and Otto Hahn [1879–1968]), initiated the first course that would usher in the coming genetic revolution just as the first atom bombs were being dropped on Hiroshima and Nagasaki. Surprisingly, this striking coincidence seems completely absent from recollections and comments about the first phage course. The first "Course on Bacteriophages" was offered at the Biological Laboratory at Cold Spring Harbor, New York, 23 July–11 August 1945. Hiroshima was bombed on 6 August, and Nagasaki was bombed on 9 August of the same year. Six students enrolled, according to the Cold Spring Harbor Laboratory records. A small but interesting class: "a marvelously motlied crowd of students," as Delbrück recalled. Motley, indeed. This first class included a first-year graduate student as well as an eminent microbiologist, president that year of the Society of American Bacteriologists. Two women and four men. Physicians and physicists.

Four of these later became household names in biological research. Phyllis Margaretten (1924–2017), age 21, was a first-year chemistry graduate student at the University of Illinois who received her M.S. degree the following year under the direction of George L. Clark (1892–1969), an early pioneer in electron optics; her thesis was "Electron Microscope Studies of Bacterial Viruses." Clark was also the mentor of the other woman in this phage class, Martha Barnes Baylor. Baylor, then age 30, received her Ph.D. at Illinois and was an early specialist in the biological applications of electron microscopy. She went on to a productive career at the University of Michigan and then at Woods Hole Marine Biology Laboratory and SUNY Stony Brook.

The men in the group were Thorbjörn Sigurgiersson, a young (age 28) physicist from Iceland, a protégé of Bohr who went on to become one of Iceland's most distinguished physicists, specializing in geomagnetism; Rollin Hotchkiss (1911–2004), age 34, an associate member of the Rockefeller Institute where he started as an assistant to Oswald Avery (1877–1955) and soon succeeded him in 1947; Herman Kalckar, age 38, a Danish-American enzymologist from the Public Health Institute of New York City who became Jim Watson's first postdoctoral mentor in Copenhagen five years later; and Stuart Mudd, the oldest in this first phage class at 52, a scion of the famous American Mudd clan, chairman of the department of bacteriology at the University of Pennsylvania, and recently elected president of the Society of American Bacteriologists (now the American Society for Microbiology). Mudd also was a prominent leader in the short-lived American-Soviet Medical Society. From this "motley crew" the famous phage course would evolve.[7]

The phage course started out with an "admission test" on the prerequisites that Delbrück set for the course: "Facility in the processes of multiplication and division of large numbers; elements of calculus; properties of exponential functions." These admission requirements often screened out classical biologists whose educations were frequently mathematically deficient. The course was planned for nine periods with nine different topics in basic phage technique, starting with preparation and plaque assay of a phage stock and isolation of a new phage from fecal material. The second period entailed microscopic observation of phage lysis and picking phage-resistant bacterial mutants. The third and sixth periods introduced serological inactivation of phages and calculation of inactivation rates and titers. The resistant bacteria were characterized in the fourth period, and three periods (the fifth, seventh, and eighth) were devoted to variations in the one-step growth experiment of Ellis and Delbrück. The ninth period was as close as the course came to exploring phage biology with the study of mixed infections and mutual exclusion effects.

As would be the custom for the next twenty-five years,

the time between periods of laboratory work was devoted to both organized lectures and informal discussion. Delbrück was famed for his preference of theorizing over direct investigation, and he seemed to instill this inclination in students whenever he had the chance. One "Delbrück story" recounted how he arranged to have clean lab glassware run in short supply once a week to force cessation of lab work to provide "time to think." Guest lectures would also contribute to the carefully planned indoctrination program of the phage acolytes.

Unfortunately, we do not have direct participant recollections of the very first phage course, but several accounts from the next few early years provide a clear understanding of the aims and successes of the phage course. One of the most enthusiastic and eloquent accounts comes from Gunther Stent (phage course 1948):

> By the time the course was over, I felt I had become an expert phagologist. I had also imbibed the conceit of the Phage Group. There was no point in paying any attention to the work of our predecessors or of contemporaries external to the "Church," as the coterie of disciples of Pope Max had been designated by the French microbiologist (and my future patron) André Lwoff [1902–1994]. Reading publications lacking the Church's imprimatur was worse than a waste of time: The unsubstantiated claims based on poorly designed experiments presented in such papers would just put wrong ideas in the True Believer's head.

Another early phage course alumnus, Seymour Benzer, referred to it as "PU," Phage University. Noel Rose (1927–2020) from the 1950 phage course also recalled the force of Delbrück, even when he was absent: "My most vivid impressions of the course relate to the personality of Delbrück himself. . . . Mark Adams ran the day-to-day laboratories and other members of the PHAGE GROUP [sic] provided lectures and led discussions. At every turn, however, there was a reference to 'wait for Max,' when a difficult theoretical problem arose. It was a bit like 'waiting for Godot.'"[8]

The role of Mark Adams, who was Delbrück's anointed stand-in and protégé, was almost uniformly recognized as the key to the long-term success of the phage course. According to Karl (Gordon) Lark (1930–2020), "He [Mark Adams] was the best teacher I ever met. He loved elegance in science and in phage he had found an object worthy of his love and colleagues that he could enjoy. . . . He straddled two worlds exemplified by the Phage Course on the one hand, and by the enzyme club on the other. Because of the work of Avery, McLeod and McCarty, he already believed that DNA was the genetic material and he also believed that phage presented a unique tool with which to investigate protein biochemistry and structure." Delbrück also paid tribute, posthumously, to Adams in his preface to Adams's book, *Bacteriophages*. "In this course were trained many of those who are presently engaged in phage research, and in addition many who are interested in related fields acquired through it a critical understanding of the phage literature. It thus served to bring phage research out of its isolation, and to foster the many links to other parts of modern biology."[9]

In addition to the memories of the course itself, participants commented on the importance of the setting. "The Phage Course did introduce me to the beauties of CSH. I spent many summers there doing research and eventually ran the Bacterial Genetics course there for several years." "The summers there [Cold Spring Harbor] for visitors was [sic] an unforgettable joy."[10]

The first year (1945) must have been considered a success, even with the small motley group of students. Demerec actively sought funds ($800) to continue the course in the summer of 1946. He succeeded enough to sustain the course, and in 1946 its enrollment doubled to twelve students, including beginners as well as seasoned senior investigators. Phage was becoming a curiosity and something to find out about. In the second and third years of the course (1946 and 1947), for example, participants included a department chair from the University of Chicago (Earl Evans, Jr.), an MIT physics professor who had helped load the bombers that carried the atom bombs to Hiroshima and Nagasaki (Philip Morrison [1915–2005]), the discoverer of the nuclear chain reaction and longtime collaborator of Albert Einstein (Leo Szilard), all struggling with phage at the lab bench beside recent college graduates Roman Kutsky (1922–2006) and Catherine Fowler (1925–2004) and newly minted Ph.D.s Harriet Taylor (1918–1968) and David Perkins (1919–2007).[11]

The phage course continued to attract acolytes until its last offering in 1970. Slightly over four hundred students are recorded in the course archives at the Cold Spring Harbor Laboratory. By then, phage research was firmly established throughout the land; Delbrück had long since lost his passion for phage in favor of *Phycomyces,* a model for neuroscience research; and many phage biologists were turning to animal virus work, spurred by the promise of federal money from the incipient "war on cancer." The grounding that these biologists received in their early work on phage, however, would lead to a fundamental change in the field of virology. The medical and pathological themes of virus research in the 1930s and 1940s, based on bioassays, tissue pathology, and production of disease, would yield to quantitative studies modeled on phage research led by former phage researchers such as Renato Dulbecco, Wolfgang (Bill) Joklik (1926–2019), and Marguerite Vogt (1913–2007).[12]

In addition to the phage course, the APG benefited from another community-building effort led by Delbrück. In the immediate postwar period, travel and rapid communication (long-distance phone calling) were not as easy as today. With a tiny and scattered collection of phage enthusiasts, Delbrück set out to bolster communication to build cohesiveness, spread his views of proper phage research, and establish his authority.

The *Phage Information Service*

Communication is one of the key features of science and, indeed, much has been written about science as "public knowledge," about the "republic of letters," and about the "invisible college." In addition to private correspondence between like-minded individuals and the formal "Proceedings" and "Transactions" of early scientific institutions such as the Royal Society and the French Academy, in the nineteenth century the "journal" evolved from private circulation of letters to a more formal mode of scientific communication, intended, as the name suggests, to communicate the daily work of scientists to a wide community of interest. Often centered on the leadership of a distinguished academic, who acted as the gatekeeper for what was to be published, by the twentieth century, journals became just as formal, as slow, and as quirky as the proceedings and transactions of the learned societies. As the philosopher of science David Hull argued, one's scientific work is of value only to the extent that it is used by others.[13]

Especially for tiny groups of like-minded researchers working in far-flung laboratories, communication was both a problem and an adhesive. Indeed, as the pace of science seemed to increase, there were "revertants" to the earlier, less formal "newsletter" forms of communication that had been especially useful to fledgling scientific constellations in the early stages of their evolution.

Given their institutional obligations and the difficulties of travel, the phage workers in Nashville, Bloomington, and St. Louis could meet face to face only sporadically, and, indeed, other even more distant researchers were further isolated. In the beginning of a nascent discipline, the community is usually not large enough to support the costs, in both financial and organizational terms, to sustain a regularly published journal. Delbrück recognized this problem for phage research, but two very successful models were already familiar to him. In the field of genetics, the *Maize Genetics Cooperation Newsletter,* started in 1929 at Cornell by Rollins A. Emerson (1873–1947) and Marcus M. Rhoades (1903–1991), provided news of current research on corn genetics as well as sources of seeds to anyone who cared to subscribe. Two institutions close to Delbrück's heart, Caltech and the Cold Spring Harbor Laboratory, had jointly produced an informal, rather sporadically published, privately circulated newsletter for the small but growing community of fruit fly geneticists. The *Drosophila Information Service,* apparently modeled on the *Maize Genetics Cooperation Newsletter,* was started in 1934 by Calvin Bridges (1889–1938) and Milislav Demerec, with the support of the Carnegie Institution of Washington. Bridges was a Carnegie fellow at Caltech in Morgan's group, the center of *Drosophila* genetics, and Demerec was at Cold Spring Harbor, in Carnegie's department of genetics. The *DIS,* as it became known, was an informal multipage newsletter, reproduced by mimeograph and distributed to a mailing list of "subscribers" at no cost.[14] There was very little review or editorial control, apparently, and often preliminary results or summary tables of data were presented, items that would not be considered suitable for publication in the regular journals. Often there were technical matters of interest to small groups of specialists. The *Drosophila Information Service* was produced (nominally) twice a year at its inception in 1934, and it is now published annually, with occasional special issues.

With such a model in mind, Delbrück initiated a parallel newsletter for phage research, first to disseminate work reported at early "phage meetings" and later to include correspondence and other results that were sent to Delbrück. In effect, this newsletter, the *Phage Information Service,* made Delbrück the de facto clearinghouse for most phage research. It also had the effect of making him the gatekeeper of the APG. Indeed, as mentioned earlier, one early phage researcher stated that the defining aspect of membership in the APG was whether or not one sent manuscripts to Delbrück for review, approval, and possible dissemination in the newsletter prior to publication.[15]

The first newsletter did not start with the first phage course or first few "phage meetings," as the informal gatherings were almost immediately christened. Issue number 1, dated March 1947, was an eight-page communication of what transpired at the "Phage Meeting, Vanderbilt University, Nashville, Tenn. February 17–19, 1947" (figure 5). The list of participants and their institutional affiliations heads the newsletter. The texts are Delbrück's summary of the talks rather than material submitted by the speakers: eight separate entries appear under the heading "Summary of Speakers, Topics, and Conclusions." Four of the eight papers (Mark Adams, Edward H. Anderson, Gus Doermann, and Delbrück) dealt with phage growth phenomena such as co-factor requirements and lysis inhibition. Two described UV radiation effects on phage (Luria and Max Zelle [1915–1980]); one focused on genetics (Hershey) and another on biochemistry (Cohen). This mix of topics with the emphasis on growth conditions and radiation effects would come to be dominant themes in the early days of the APG.

Phage Meeting, Vanderbilt University, Nashville, Tenn.
February 17-19, 1947

Participants

M. H. Adams, Dept. Bact., New York University, College of Medicine
E. H. Anderson, Monsanto Chem. Co., Oak Ridge, Tenn.
T. F. Anderson, Johnson Fdtn., Univ. of Penn., Med. School
S. S. Cohen, Dept. Pediatrics, Univ. of Penn., Med. School
A. D. Hershey, Dept. Bact. and Immun., Washington Univ., Med. School
S. E. Luria, Dept. Bact., Indiana Univ.
F. Putnam, Dept. Biochem., Univ. of Chicago
M. R. Zelle, Natl. Inst. of Health, Bethesda, Md.

-W. T. Bailey, Jr.   )   *Demerec*
-M. T. Bush   )   *Newcombe*
-M. Delbrück   )   *Monod*
-A. H. Doermann   )   Vanderbilt University   *Laterjet*
-F. Phillips   )   *Beadle*
-N. Underwood   )   *Beard*
                                                          Pátau

Summary of Speakers, Topics, and Conclusions *Ingersoll*
                                                          Sharp

A. H. Doermann: "On Lysis Inhibition"

Lysis Inhibition occurs in cultures infected with $T2r^+$, $T4r^+$, $T6r^+$/
It is caused by an inhibitor released from the first bacteria which
lyse. This inhibitor delays the lysis of the remaining bacteria
for varying lengths of time, generally around 20-60 minutes. This
inhibitor is shown to be phage itself by three methods:
1. It is thrown down in the centrifuge approximately at the
   same rate as is phage. None of the inhibitor can be dia-
   lyzed. Purified phage inhibits as effectively as crude
   lysates.
2. It is inactivated by specific antiserum at the same rate
   as is the infectivity of the phage.
3. It can be absorbed from lysates by those, and only by
   those, bacterial strains which specifically absorb the
   phage.
Inhibition, then, is due to a second infection by an r^+ particle.
It occurs only if the first infection, too, is due to an r^+
particle, but there is cross inhibition between T2, T4, and T6.

There is some indication that the bursts of lysis-inhibited bacteria
are up to twice as large (as to phage yield), as those of non-
inhibited bacteria.

Nephelometric measurements show a peculiar __diminution__ in scattering
power of bacterial cultures during the first 8-10 minutes after
addition of phage. Part of this "first dip" may be due to "Lysis

Figure 5. First issue of the *Phage Information Service*, page 1
(Courtesy of the Caltech Archives)

Both themes in this early phage meeting reflect Delbrück's background and interests. As has been noted, Delbrück's approach to phage reproduction was to consider the system inside the cell as a black box to be studied indirectly by manipulating the external conditions and measuring the output results.[16] Medium, co-factor, and radiation effects were clear and simple experiments to be performed with little ambiguity. Likewise, the output—the phage yield—was an unambiguous endpoint. This indirect approach to the black box of the phage-infected bacterium was precisely that of the experimental atomic physicists of Delbrück's youth. The one paper on biochemistry by Seymour Cohen represented a continuing tension that developed in the APG, a tension that at times was deliberately overlooked but was there, nonetheless. For the physicist, new to biology, the biochemical approach (often rather pejoratively termed "grind and find") required messy methods that were often qualitative rather than quantitative and that required more specialized expertise beyond simple models, a critical mind, and clear thinking.

The first issue of the *Phage Information Service* was circulated to the scientists who took part in this phage meeting as well as others who could not or did not attend.

Newsletter recipients who participated in the 1947 meeting
Mark H. Adams (New York University)
E. H. Anderson (Monsanto, Oak Ridge)
Thomas F. Anderson (University of Pennsylvania)
W. T. Bailey, Jr. (Vanderbilt University)
M. T. Bush (Vanderbilt University)
Seymour S. Cohen (University of Pennsylvania)
Max Delbrück (Vanderbilt University)
Augustus H. Doermann (Vanderbilt University)
Alfred D. Hershey (Washington University, St. Louis)

Salvador E. Luria (Indiana University)
F. Phillips (Vanderbilt University)
Frank Putnam (University of Chicago)
N. Underwood (Vanderbilt University)
Max R. Zelle (National Institutes of Health)

Newsletter recipients who did not participate in the 1947 meeting
George Beadle (Caltech)
Joseph Beard (Duke University)
Milislav Demerec (Cold Spring Harbor Laboratory)
? Ingersoll
Raymond Latarjet (Radium Institute)
Jacques Monod (Pasteur Institute)
Howard B. Newcombe (Chalk River, Canada)
Klaus Pätau (Kaiser Wilhelm Institute Berlin)
? Sharp

The second issue of the *Phage Information Service* (number 2, March 1948) was distributed more widely. Delbrück's distribution list, noted in the margin of his copy of the newsletter, has twenty-four names, which included most of the recipients of issue number 1. Missing, however, were Zelle, Putnam, E. H. Anderson, and several of Delbrück's Vanderbilt associates. By the time of this issue, Delbrück had moved from the physics department of Vanderbilt University to Caltech, as a professor of biology. This newsletter provided a nine-page summary of Delbrück's discussion and seminar notes on the visits to Caltech of Salvador Luria and Thomas Anderson in February 1948. Thus, in addition to providing the dispersed phage community with meeting proceedings, Delbrück was making Caltech—and the people who visited him there—the center of knowledge distribution of a more informal kind. Like the nineteenth-century journals that evolved from the correspondence and communication with eminent European professors into established journals, Delbrück's newsletter was adding to his own construction of the new discipline of phage research, and it was beginning to define the scientific community that would become the APG.

An indication of Delbrück's longer-range vision for phage and viruses was his attempt in 1950 to foster a grand synthesis of virological knowledge from basic laboratory studies to clinical medicine spanning all of virology, from microbes to plants to animals. He organized a conference at Caltech in March 1950 with the sweepingly inclusive title "Viruses 1950" to bring together the luminaries of mid-century virology to see what common themes could be identified (figure 6). A critical look at the invitees and topics suggests more than a little bias toward making virology over in the image of bacteriophage. This conference, which resulted in a small volume that was important to phage workers at least, took place just before the watershed discoveries about DNA structure and the nature of the gene, and for that reason alone it was unfortunately premature.[17] The conference and the publication, however, served Delbrück's program to promote phage as central to the field of virology and to establish Caltech as the center of cutting-edge phage work.

The last newsletter in the Caltech archives is issue number 14, which is devoted to the phage meeting at Cold Spring Harbor Laboratory held 24–26 August 1959. It appears to contain the verbatim abstracts submitted by ten of the participants. In this same year, Delbrück left Caltech to spend several years in Cologne as the initial director of the Institute for Genetics; this institutional move, along with the shift in his research

Figure 6. Participants at "Viruses 1950" conference at Caltech,
20–22 March 1950 (Courtesy of the Caltech Archives)

away from phage to the photo responses of the fungus *Phycomyces,* contributed to the demise of his personal project, the *Phage Information Service.*

Although the phage newsletter had been absent from the scene for over a year, phage biologists still saw the need to continue certain formal communications. One need related to the network of informal exchanges of strains and mutant forms of phages and their hosts. In the wider world of genetics, stock centers and records of biological lineages had been maintained for some time. The Genetics Society of America had a standing committee devoted to the maintenance of genetic stocks, and the American Type Culture Collection (ATCC), established in 1925, was a major custodian of certain microbial stocks. After the Cold Spring Harbor Laboratory phage meeting in the late summer of 1960, a subcommittee on bacteria and phages met to assess the need for a new stock center devoted to phages and their hosts. Chaired by Vernon Bryson (1913–1985) of Rutgers University, the subcommittee was made up of Louis S. Baron, Demerec, Doermann, Sol H. Goodgal (b. 1921), and Emilio Weiss (1918–2008), the latter two being absent. Although this subcommittee did not see the need for an additional stock center to maintain and distribute phage stocks, believing that the existing culture of lab-to-lab sharing was working well, its members did agree on the need for an annual updated list of phages held by various labs. They proposed the creation of a Phage Stock List to be overseen by a volunteer editor.

Gunther Stent was approached to take on this task. Bryson, the chairman, wrote to him: "It seems that the Virus Laboratory [at the University of California, Berkeley] might have enough secretarial and mimeographing help to put together one bulletin a year, under the general supervision of you and your various colleagues. Since the lists are to be by categories, rather than by itemization of each stock, even a big collection of several thousand mutants should take only a few pages. The responsibility of having things right is on the contributor, not on the editor."[18] Bryson then held out the inducement that the Genetics Society might provide modest funds for an assistant, should Stent accept the editorship. No response to this proposal is to be found among Stent's archival records, and no examples of a Phage Stock List are known. Perhaps the APG had become sufficiently mature to thrive without continuation of such discipline-building efforts.

Challenges and Accommodations

To some extent the legitimacy of the APG would depend on adherence to the shared commitments and community values established by common agreement via phage course indoctrination and *Phage Information Service* articles. Nonetheless, there were controversies and challenges that had to be confronted. Lysogeny, as will be discussed later, was one major challenge that was eventually absorbed into the fold of legitimacy. Other cases involved the discovery of phages that had properties that were outside the range of the standard T-phages, in particular phages that had either single-strand DNA genomes or RNA genomes. Furthermore, since the *E. coli* host of the T-phages is a gram-negative bacterium, it has certain experimental disadvantages that are not exhibited by gram-positive organisms such as *Bacillus subtilis* and *Bacillus megaterium:* gram-positive bacteria can be permeabilized so that they will take up and incorporate added naked DNA in a process dubbed "transformation." (Only years later were methods discovered to permeabilize gram-negative bacteria such as *E. coli.*) Phages that infected *Bacillus* species (φ29, PBS-1, and many temperate phages) were grudgingly accepted into the pantheon of "other phages" along with the small single-strand DNA phages such as φX174, fd, and M13, and the RNA genome phages such as f2, R17, MS2, and Qβ.

In the early 1950s, with crucial advances in the understanding of the chemistry of the gene—that is, its identity as nucleoprotein, the double helical model of DNA, and beginning biochemical studies on nucleic acid metabolism—the original APG research framework and its shared elements had to adapt to a more chemical approach: more biochemistry, less radiobiology, and less reliance on theoretical models and reasoning. Reproduction, in the biological sense, gave way to replication, in the chemical sense, but the basic element of under-standing heredity continuity in mechanistic terms could be retained. Phage biology was adaptable to new ways of exper-imentation while still retaining the fundamental principles of quantitative science, another key element of the framework of shared commitments of the original phage group. In a way, the very success of the initial framework of shared commitments would produce results that would require refinement and am-plification as new information appeared and new discoveries were made. But keeping the APG "pure" became more and more difficult and harder to defend.

8

Place in Science
Cold Spring Harbor from
Quantitative Biology to Phage

On 9 December 1946, back home in Paris, the young French biologist Raymond Latarjet wrote to his recent post-doctoral mentor, Milislav Demerec, director of the Cold Spring Harbor Laboratory in New York:

10,30 − 6 = 4,30 P.M. This is about the time for the Staff meeting. Miss Carlery just brought the tea on the table, and Mrs. Smith is looking for the cookies. As it rains, Barbara is putting on her overshoes, ready to cross from the animal house without getting dirty her pant no. 372[?]. Bryson's Buic [*sic*] stopped along the northern wall of the main building; its owner prepares the next joke he will utter in the most serious manner. Dr. MacDowell is already at work, staring at the tea over his glasses, and carefully balancing its darkness with boiling

water. Evelyn [Witkin] enters the room silently, glid-
ing to her seat; she puts before her on the table an
ash tray and a cigarettes package—none but Phillip Morris—. On the other side of the table, Dr.
MacDonald jumps faster to her seat, handling her
cigarettes—none but Raleigh (tipped)—after a tour
of the library, Kauffman comes, smiling, and, while
Dr. MacDowell starts asking "medium? light" Dr.
Demerec arrives and takes the last chair.

Are you as many as we were last year? Who
did replace all these sparkling minds, Italian, Chil-
ian, or French, who left? Is everything the same despite the coal strike, the republican shift and Byrd's
expedition to the South pole? I am pretty sure it is,
and very often I miss the calm good mood which
pervades Long Island's shore and which provides
such a good yield to work.[1]

Two decades later James Watson recalled his graduate
student days at Cold Spring Harbor: "I looked forward greatly
to the forthcoming summer (1948) when Dulbecco and I would
go with the Lurias to Cold Spring Harbor. Delbrück and his
wife Manny were coming for the second half while, before
they arrived, there was to be the Phage Course given by Mark
Adams. . . . As the summer passed on I liked Cold Spring Harbor more and more, both for its intrinsic beauty and for the
honest ways in which good and bad science got sorted out."
And Salvador Luria, another APG leader and Watson's dissertation adviser, enthused: "Cold Spring Harbor was a perfect
setting, for fun and for work. There for many years we spent
summers working and teaching—on bacteriophage at first,
then on other exciting topics in molecular biology."[2]

Gunther Stent, the self-appointed historian of the APG, described his introduction to the Cold Spring Harbor Laboratory in his autobiography:

> Sensing the atmospheric electricity of a high-brow dolce vita, I instantly fell in love with the place.
>
> The lab assigned me a single room on the second floor of Blackford Hall. Blackford was one of the first reinforced-concrete structures built on the Eastern Seaboard and, as far as I could tell, had had no maintenance work in the forty years since its erection. Blackford was the lab's main social center. Its ground floor housed the kitchen and cafeteria on the west side and a lounge on the east side. Even though I was a hardened six-year veteran of institutional food, I found the three daily meals served in the cafeteria as awful as they were skimpy. The crummy lounge was furnished with dilapidated sofas and easy chairs that the Salvation Army would have refused as a charitable donation. It was used as a parlor, as a seminar room and, incredible as it seemed, as the venue of the world-famous annual Cold Spring Harbor Symposium on Quantitative Biology.
>
> The second floor of Blackford was divided into a couple of dozen cell-like single rooms for senior (postdoctoral) gentlemen summer visitors, who did their ablutions in a communal, minimalist toilet/shower facility. Blackford also had a basement, whose layout suggested that it once served some social functions. It was now under water most of the time.[3]

As the editors of the Delbrück festschrift in 1966, *Phage and the Origins of Molecular Biology*, noted: "Two scientific institutions keep recurring in the following pages—the California Institute of Technology and the Cold Spring Harbor Laboratory. These two places, at opposite ends of the American continent, were the Mecca and Medina of the Phage Group to which the faithful made their periodic *hadj*. . . . Caltech, though not large by the standards of American universities, is a powerful and world-famous institution. . . . But why Cold Spring Harbor, a small station of meager resources on the shores of Long Island Sound, should have played such an important role is not immediately obvious."[4]

A young geneticist, Allan Campbell (1929–2018), who would go on to solve a major mystery of lysogeny (to be described later), recalled his situation in 1957: "I was not offered any position in arts colleges at that time, but Milislav Demerec, Director of the Carnegie Institution Department of Genetics at Cold Spring Harbor Laboratory, wanted a one-year appointment to replace George Streisinger [1927–1984], who was spending the year abroad. Cold Spring Harbor Laboratory was a marvelous place to do research, free from other obligations, and the senior people—Demerec, Barbara McClintock [1902–1992] and Al Hershey—provided inspiration and moral support."[5]

For almost every member of the APG, the laboratory at Cold Spring Harbor, New York, became a place of reverence and scientific excitement. A place of myths and memories. A place that the fledgling phage biologists felt was their intellectual home. What made this place, this laboratory, special? How did its iconic status as the "Medina" of bacteriophage develop?[6]

Place, at its most local example, is the individual laboratory, a place where, in the words of historian F. L. Holmes, reality is "elaborated," where students are indoctrinated, where publications are produced, and "facts" are assembled. At a higher level, place includes venues for meetings, such as the Royal Society, the *Académie des sciences,* or perhaps sites of material collections such as gardens, museums, and libraries. Other places are important as meeting venues that have developed individual cultures and norms: the Gordon Research Conferences (for scientists), the Solvay Conferences (for physicists and chemists); and the "Kalamazoo meeting" (for medievalists). At the more conceptual level, place has been applied to the location of scientific activity, not quite attached to a specific piece of real estate, but geographical nonetheless: the Paris Clinic (medicine), the Göttingen School (history); the Vienna Circle (philosophy); the Chicago School (economics). More institutional and functional is the notion of place as a facility for research such as CERN, Los Alamos, the National Institutes of Health, or even the International Space Station. Place has even come to stand for unique scientific or other kinds of historic events: Asilomar (gene cloning); Alamogordo (Trinity atom bomb test); Bretton Woods (post–World War II world economic planning); Yalta (post–World War II geopolitical planning).

Although Von Eckardt's analysis did not take into account *place* as a component of discipline formation, there are good reasons to include it in an account of factors that contribute to the coherence and stability of an intellectual enterprise. Place provides a substratification of specific communities within the larger society, a way to fix and perpetuate such communities the same way a hallowed site provides identity to the religious as a destination for pilgrimage and ritual. Science has its rituals, too, and they are strengthened, fixed, and transmitted by association with place. While the "invisible college" or its early precursor, the "Republic of Letters," represents the idea of community within a larger society, the addition of place to community identity allows for consolidation of personal relationships and negotiation of agreement on community norms (accepted problems, methods, and evidentiary standards). In the early stages of discipline formation, such as that of molecular biology, community identity, helped by ritual, history, and place, was needed to establish the scientific and social parameters that delineate the new discipline from the old, that provide the framework for community entry and participation, the emergence of authority and leadership, and recruitment and education of the novice. Last, but not least, place is the locus of specific knowledge production and circulation.

One category of place that has been important in biology has been the "experiment station," a place located strategically close to relevant biological material required for study of a specific set of biological problems. Places such as Naples, Italy (Anton Dohrn Station), Pacific Grove, California (Hopkins Marine Station), and Woods Hole (Massachusetts) Marine Biology Laboratory have functioned as a combination of traditional laboratory, meeting venue, and source of biological specimens.

Place in science often calls to mind crucial activities such as communication and "networking," development of collaborations, resolution of scientific controversies, mentoring, career planning, and the social life essential to all these other activities.

How, then, can we understand the evolution of the Cold Spring Harbor Laboratory as the center of APG identity? Science, like all human activities, is influenced both by physical and social environments. The physical surroundings such as laboratories, living quarters, climate, and geography interact with social factors such as size and composition of the intellectual community, characteristics of leadership, group attitudes and values, and local and national identities. Historians of science have analyzed such factors as nationality, intellectual leadership, and gender as key determinants affecting the scientific enterprise. Besides these macro-level influences, strong local cultures evolve that have clearly differentiated certain places into centers for one kind of science or another. Studies of "the laboratory" and of "research schools" have been especially fruitful in understanding the complexities of science as organized social structures.[7]

A particular scientific field or scientific approach is often associated with a particular place as well as specific people, and phage research has been no exception. In addition to the academic centers of phage research in North America—Caltech, the University of Chicago, Indiana University, and a few other schools—the Cold Spring Harbor Laboratory on Long Island, New York, takes pride of place for the APG (figure 7). This was a special place with a special meaning for phage workers from the very start. Not only was this independent laboratory the site of education and proselytizing for phage biology, but also it was a common meeting ground for the far-flung members of the tiny APG, working, often in total isolation, dispersed in the large and small academic institutions across America. It is a place where new phage workers gave their first papers, where established scientists nurtured future generations of phage workers, where social and scientific relationships were built, and where origin stories and myths were developed, all contributing to the growth and cohesion of the newly developing field.

But how did the Cold Spring Harbor Laboratory come to play this key role? The evolution of Cold Spring Harbor Laboratory as the spiritual home of the APG was the result of both conscious planning and fortuitous events. Before the current de facto year-round mélange of academic life involving both teaching and research modes, research was something many

Figure 7. Cold Spring Harbor Laboratory, circa mid-twentieth century: Carnegie Laboratory, covered in ivy; and panorama from Long Island Sound (Courtesy of Cold Spring Harbor Laboratory Archives, New York)

college faculty scientists could pursue intensively only during the summer months when teaching duties were light or non-existent. Colleges and universities frequently did not invest in facilities for faculty research, so it was traditional, especially for biologists, to visit "research stations" where full-time research was possible for a few months. Because biology was to a large extent organismal, these stations were usually located near the sources of fresh biological material, often by the sea for ready access to marine specimens. Well-off professors initially decamped for the summer with their families, and sometimes their students, to seacoast locations where marine research could be conducted in often pleasant surroundings with primitive, makeshift laboratories. Soon, however, more permanent arrangements evolved, and Louis Agassiz (1807–1873), who with his students had dragged and dredged up marine fauna along the New England coast for several years, in 1873 founded the Anderson School of Natural History at Penikese, an island situated at the mouth of Buzzard's Bay in Massachusetts. Although this effort did not last much past Agassiz's death later that year, it has been credited as the start of the modern "research station" movement.[8]

This tradition of the scientific station grew during the late nineteenth century, strongly influenced by the success of marine biology laboratories located at useful places by the sea where biologists interested in marine organisms for physiological, morphological, and economic purposes could visit, conduct research, collaborate, and teach for short periods of time. Some of these stations became favorite places for key scientists who came to dominate the scientific culture of the station. The local availability of certain marine organisms also contributed to these individual research cultures. One of many but probably the most famous and successful in Europe was the zoological station in Naples (*Stazione Zoologica Anton Dohrn*). In the

United States, biological stations were founded (among other places) at Woods Hole on Cape Cod in Massachusetts and at Cold Spring Harbor in New York. At these places, there was a mix of educational summer courses, scientific conferences, and research opportunities. During the winter, they had only skeleton staff; the development of full-time research did not take hold until well into the twentieth century.

In the United States, the laboratories at Woods Hole and Cold Spring Harbor were in some ways friendly competitors since they were both located on the North Atlantic coast where material for marine biology was readily at hand. These were laboratories that depended on local and national patronage, and both had their periods of self-reflective uncertainty, financial crises, and debates about the future of experimental biology. Both were located near the rich and powerful in American society and relied on their patronage for financial and practical support. These laboratories were closely associated with prestigious American universities where their directors and staff held professorships. By the 1930s, however, each laboratory seemed to have evolved or, perhaps more accurately, proactively developed its own character in terms of research. This development was the product of the visions and decisions of individual laboratory directors as well as the particular scientific progress that became associated with the laboratory. Very roughly, one might characterize Woods Hole as a center of embryology and developmental biology, while the Cold Spring Harbor Laboratory became a center of genetic biology.

The history of the Cold Spring Harbor Laboratory from the mid-1930s provides insights into its becoming the home of phage biology and its central role in the life of the APG. In a nutshell, this history parallels that of molecular biology writ large: from biophysics in the 1920s and early 1930s to radiation biology in the later 1930s and early 1940s to genetics in the 1950s

and beyond. The Cold Spring Harbor Laboratory was born out
of the late nineteenth-century scientific optimism and philanthropic interest in marine biology. Following the model of the
Anton Dohrn research station on the bay of Naples, founded
in 1872, and the Marine Biology Laboratory at Woods Hole,
Massachusetts, founded in 1888, the Biological Laboratory was
founded at Cold Spring Harbor in 1890 by an amalgam of aca-
demics led by the Brooklyn Institute of Arts and Sciences and
wealthy New York philanthropists. Its initial goal was to provide both laboratory and field experience to college students
and teachers in natural history and aquatic and marine biology through summer courses. Its second director, Charles B.
Davenport (1866–1944), appointed in 1897, planted the seeds
of its later development with his vision of "quantitative" biology. Davenport, who held a professorship at the University of
Chicago while serving also as the director of the laboratory,
would go on to fame (or notoriety) as one of the key founders
of the eugenics movement in the United States. In the summer
of 1903, another key event that presaged the future of the Cold
Spring Harbor Laboratory took place: the newly formed Carnegie philanthropy, the Carnegie Institution of Washington,
established its department of experimental biology and funded
a Station for Experimental Evolution at Cold Spring Harbor on
land leased from the organization that owned the Biological
Laboratory. These two laboratories, existing side by side, often
so intertwined that they seemed like one, provided both a critical mass of science and a creative tension that existed until 1961
when the Carnegie unit and the remnants of the Biological
Laboratory were merged and resurrected as the Cold Spring
Harbor Laboratory of Quantitative Biology.[9]

Ever since the work of Gregor Mendel, educated in com-
binatoric mathematics, the study of heredity took a quantitative
bent, a direction that was diversified a bit later by the biometric

approaches to genetics of Francis Galton (1822–1911). It is not
so surprising, then, that at the Cold Spring Harbor Laboratory
"quantitative biology" came to mean "genetics." Davenport's
eugenics studies, conducted under the rubric of experimental
evolution, were quantitative but not very experimental. Other
work in the Carnegie-sponsored department involved the genetic study of model organisms, fruit flies (*Drosophila*), maize
(corn, in American usage), and fungi. The young maize researcher Milislav Demerec joined the Carnegie genetics group
in 1924 and soon switched his interest to fruit fly genetics.
Later, he would play a profound role in the eventual development of the Cold Spring Harbor Laboratory as the intellectual
home of phage and microbial genetics. In the 1920s, however,
the summer courses at the Biological Laboratory were not
enough to maintain the health of the institution, and in the
mid-1920s, Davenport looked to the new field of "biophysics"
as a way for the Biological Laboratory to reinvent itself. Biophysics at this time was an amalgam of quantitative physiology and new laboratory techniques based more on physical
methods than on older low-tech approaches. One of the newer
techniques was the use of ionizing radiation to perturb cell
function, in particular, the "target theory" as described earlier.

In 1924, Reginald Harris (1898–1936), a biologist who had
worked with the mouse geneticist Clarence C. Little (1888–1971)
and who had studied ecology and evolution in native populations in South America, became the director of the newly
reorganized Biological Laboratory under the care of the Long
Island Biological Association. Harris, like his father-in-law,
Davenport, who continued as director of the Carnegie unit,
was enamored with the possibilities of "biophysics" and formed
an Advisory Committee for Physiology and Biophysics. In
1928, the first recruit in this new area was Hugo Fricke (1892–
1972), a former protégé of Niels Bohr and by then a physiolo-

gist working on the effects of x-rays on the physical properties of the cell membrane. Fricke and several collaborators published a series of papers in this general field, but his renown, by far, rested on his development in 1925 of a chemical method for quantitative radiation dosimetry, which still bears his name.[10] This method, based on the oxidation of ferrous ions to ferric ions by the x-ray products in aqueous solution, provided a simple, accurate, and reliable chemical dosimeter that revolutionized studies in radiation biology. In 1955 Fricke moved to Argonne National Laboratory, where he continued his work on radiation chemistry.

By the end of the 1920s, Demerec had established the Carnegie unit at Cold Spring Harbor Laboratory as one of the three centers of *Drosophila* genetic work in the United States, along with T. H. Morgan at Columbia and Hermann J. Muller in Texas. As a result of Muller's Nobel Prize–winning discovery of the mutagenicity of x-rays, genetics and radiation had become crucial partners. Debates on the physical nature of "the gene" relied heavily on understanding radiation mutagenesis.

Demerec's interest in mutagenesis, first in fruit flies and later in bacteria, led him to the thriving field of radiation biology and biophysics; he was a recognized authority in *Drosophila* genetics as well as radiation biology more broadly. The focus on radiation studies was central to understanding the nature of the gene and the process of mutagenesis. Demerec had organized conferences on the nature of the gene that met in Woods Hole, first in 1936, then in 1938, and finally in the fall of 1940, prior to any hint of phage work. Among the stated aims of these "Gene Conferences" were to "coordinate work on gene problems," to discuss unpublished material "to speed up the tempo of the work," and "to bring together for informal discussions a group of workers actively interested in the gene problem in its broadest sense." Clearly, in this aspirational language, the seeds of the Cold Spring Harbor Laboratory phage meetings can be noted. When the opportunity arose to support the tiny group of phage biologists, Demerec envisaged the Cold Spring Harbor Laboratory as a center for this new field, just as Woods Hole had come to represent embryology. He encouraged a broad range of work devoted to quantitative genetic studies. The Cold Spring Harbor Laboratory was the site of fledgling work by Claude Shannon (1916–2001), later a founder of communications theory, and Esther Zimmer, who later discovered phage λ and who, with her husband, Joshua Lederberg, clarified a basic conundrum of evolution with her replica plating experiments on bacterial mutants.[11]

An integral relationship developed between the Biological Laboratory at Cold Spring Harbor and the Carnegie Institution of Washington as a result of the location of Carnegie's department of genetics on the same site in Cold Spring Harbor. This connection added more support for the biophysical flavor of research there. The Carnegie Institution of Washington—founded in 1902 by the philanthropist Andrew Carnegie as "an institution which . . . shall in the broadest and most liberal manner encourage investigation, research, and discovery [and] show the application of knowledge to the improvement of mankind"—sponsored an eclectic group of researchers but seemed to favor the physical sciences and problems related to evolution and (initially) eugenics. In 1946, the physicist Merle A. Tuve (1901–1982), director of the oddly named "Department of Terrestrial Magnetism" of the Carnegie Institution, convened the Ninth Washington Theoretical Physics Conference on the subject of "The Physics of Living Matter," which invited the luminaries of physics at the time, including Bohr, Szilard, John von Neumann (1903–1957), Edward Teller (1908–2003), Franck,

and Gamow, as well as several dozen geneticists such as Demerec, Muller, Beadle, Stanley, and Hollaender.[12] "Living Matter" seemed to mean "the Gene."

It was natural, then, for the young Max Delbrück, having just collaborated with the radiation geneticist Timoféef-
Ressovsky on their soon-to-be famous "Three-men-paper," to
visit Demerec at the Cold Spring Harbor Laboratory in the fall
of 1937 on his way from Germany to Caltech. Delbrück spent
a rather depressing month (as he later described it) with Demerec at Cold Spring Harbor trying to learn the laboratory
skills of fruit fly cytogenetics.[13] The second convergence on
the Cold Spring Harbor Laboratory came from Salvador Luria,
who had arrived at the College of Physicians and Surgeons
(Columbia University) to join the laboratory of the biophysicist Frank Exner. Columbia, along with several other academic
institutions, had loose ties to Cold Spring Harbor Laboratory,
and Demerec had just recruited the émigré physicist Ugo Fano
to his biophysics team. Fano, as it turned out, was a childhood
friend of Luria's from Turin. Consequently, Cold Spring Harbor Laboratory became part of Luria's intellectual and personal
orbit by dumb luck.

The meeting of Luria and Delbrück is the stuff of legend
(a fuzzy legend, at that), but both had connections at Cold
Spring Harbor Laboratory, so when they were planning a summer research collaboration in 1941, Cold Spring Harbor Laboratory, near to Luria in New York, familiar to Delbrück, and
with available summer laboratory space, became their place
of choice for their first real work together. Delbrück's initial
rather gloomy views on the Cold Spring Harbor Laboratory,
developed during his late autumn adventures in fruit fly cytogenetics in 1937, gave way to a more sunny impression of the
laboratory in summer. Luria recounted their first summer there,
enhanced by Delbrück's bride, Manny, as "truly a movable feast

... a perfect setting for fun and work." Luria credits Demerec and his revitalization of the laboratory with providing the right kind of environment hospitable to summer visitors excited about science.[14]

In Luria's description, life at Cold Spring Harbor involved monastic living quarters, with a decrepit physical plant but a spectacular setting on a small bay of Long Island Sound, with its own swimming beach for midnight dips, a protected harbor for small sailboats owned by some of the more affluent researchers, and a sense of community impressed on the denizens of the laboratory by both its seclusion and its genteel austerity. Phage research turned out to be a good fit for such an environment. It did not require much in the way of specialized equipment: the basics of a microbiology laboratory were all that was needed. An autoclave, an incubator, petri plates, pipettes, test tubes, and lots of flat table space would do nicely for research that was based on imaginative permutations and combinations of growth conditions and plaque counts. In contrast, too, with the more traditional work of marine biology, phage research was not at the mercy of erratic harvests of local material such as squid, sea urchins, and the like. It was fast, too. Plaques formed in hours. An experiment done in the morning could be analyzed and repeated by nightfall. Indeed, research was often so intense that the supply of sterile pipettes was exhausted by Monday morning when the lab's glassware washers returned from their weekend off. Delbrück was reputed to have declared Mondays as a day for thinking and talking rather than for more experiments, while they waited for replenishment of the sterile pipette supply. This enforced downtime was, no doubt, his way of encouraging reflection rather than his belief that scientists could not, if needed, wash glassware.[15]

As Demerec sought to "reinvigorate" the Cold Spring Harbor Laboratory along the lines of biophysics, he was also

faced with financial troubles.[16] The summer course was one of
several ways that he envisioned to keep the laboratory full, en-
hance its reputation, and bring in some badly needed funds.
His goals coincided with the idea, attributed to Luria but put
into practice by Delbrück, to recruit people to the new field of
phage biology by discipline building with a course that would
educate new and experienced scientists into the mysteries of
phage but also, even more importantly, indoctrinate them according to the vision of Delbrück, Luria, and their approved
acolytes. Thus, starting in 1945, Cold Spring Harbor Laboratory became the locus of the "Phage Course," which soon became associated with an annual conference at the end of the
course, an "alumni reunion" of former students of the course,
now active phage workers. Cold Spring Harbor Laboratory, in
the manner of other educational institutions, would thus develop a culture, ethos, and mythology. It was an alma mater to
many phage workers, truly a "fostering mother." Like all such
institutions, lore, myths, and rituals developed that were passed
down from older to younger. Delbrück, especially, was sensitive to these aspects of discipline building. His early experiences in a large, upper-class German academic family gave him
the background to provide just the right mix of authoritarian
and iconoclastic confidence to assume the leadership of this
evolving discipline. A staple of phage meetings was always the
retelling, often around a late-night bonfire on the lab beach, of
"Delbrück stories." Rarely there were "Hershey stories," never
"Luria stories."

The Cold Spring Harbor Laboratory and the APG became symbiotic, their identities and fates inextricably linked.
This relationship was certainly fostered by the affinity of key
members of the phage community for the place where they
would spend at least part of their summer holiday, an affinity
reinforced by Demerec's welcoming attitude toward these vis-

itors. Delbrück, Luria, Watson, and the staff of the phage course were frequently "in residence," making the place an important stopover for European visitors arriving at the main entry point on the East Coast, one of the two New York airports. From 1950, too, Al Hershey was a permanent fixture at Cold Spring Harbor Laboratory as a staff member of the Carnegie Institution laboratory stationed there.[17]

The fact that Cold Spring Harbor Laboratory became the home of the famous phage course (as the Hopkins Marine Station in Pacific Grove, California, would become associated with C. B. van Niel's general microbiology course from 1938 to 1962) provided the place with an ever growing list of alumni, many of whom took up phage research and then sent their students to the phage course and "summer camp" at the Cold Spring Harbor Laboratory. The alumni of the phage course, together with others who regularly attended the short, informal meetings that were the culmination of the several weeks of the course, formed a community of both shared scientific interest and shared social experiences. Neptune's Cave, the local pub a short walk from the laboratory grounds, late-night swimming off the sand spit that juts out into the harbor, and perennial jokes about the cafeteria food were common experiences shared by generations of phage workers who thought of the laboratory as intrinsic to their scientific identity. The cohesion and association of the APG with the Cold Spring Harbor Laboratory was far more pronounced than the identification of other research traditions with other research stations. This close bond of work and place was the culmination of the successful discipline-building strategies of Delbrück and Demerec.

A decade after the first phage course, it was clear that Demerec viewed his institution as the center of the APG. In the summer of 1954, he discussed his vision for a "summer center for phage workers" with key members of the group, Benzer,

Doermann, Levinthal, Stent, and Watson. In a letter to this group in the fall of 1954 he sought their input on the possibility of "reorganizing our summer laboratories so that they would be suitable and attractive for phage workers," a place "where the group could get together for conferences and discussions and also where adequate facilities would be available for research." Demerec suggested that, if he had reasonable assurances that the group would be sufficiently large and that they would return for a reasonable number of summers, he would seek funds to upgrade the laboratories with the newer expensive equipment such as centrifuges, radioactive counters, and the like. Apparently, he received the responses he sought, and in February 1955 he notified Benzer, Stent, and Watson that he would have renovated and equipped labs available for the summer of 1955 and that several phage workers—including the summer visitors Doermann and Luria as well as the full-time researchers Hershey and Vernon Bryson—would be in residence both for the phage course (with Doermann in charge) and summer research work. Demerec proposed several arrangements for sharing (and supporting) major equipment as well as services of support staff. The Cold Spring Harbor Laboratory had clearly staked its claim as the home of the APG.[18]

It was not geographic determination but rather a mix of contingent factors such as people, patronage, and circumstance that made the Cold Spring Harbor Laboratory both the national gathering place for the APG and the intellectual home for many other molecular biologists. Demerec's vision to develop biophysics (meaning radiation biology) as an area of emphasis in the 1930s and his recruitment of physicists such as Fricke and Fano, even prior to any thoughts of phage research, made Cold Spring Harbor Laboratory a place that could be welcoming to the new crop of physicists doing biology during and immediately after World War II. That Emory Ellis could not continue phage work, thus encouraging Delbrück to seek out Luria, and that Luria had made contacts at Cold Spring Harbor Laboratory through Fano, his childhood friend from Turin, were contingencies that simply happened without necessity. The resonances between Delbrück and Luria and their complementing personalities, together with shared research aims, worked to exploit the environment at Cold Spring Harbor Laboratory to attract and support the small, scattered community of like-minded phage researchers across the country. Communication, patronage, and indoctrination were discipline-building essentials that depended on a congenial and welcoming physical place; in the 1940s and 1950s, especially, the Cold Spring Harbor Laboratory filled the bill perfectly, and Milislav Demerec's drive and vision made it happen.

9

The Challenge of Lysogeny
Pro and Con

A viable framework for a successful research community must have both stability and resilience. It must have the commitment of its users to guide their work so that the community has coherence, yet it must have sufficient adaptability to accommodate the inevitable new theories and new facts that research generates. The extent to which a research framework successfully responds to challenges to its fundamental shared commitments—reinvigorating itself with revised elements and new versions to deal with new knowledge—shows how effective and robust the research framework is. One such challenge to the APG's framework of shared commitments is represented by the problem of lysogeny.

Almost from the first discovery of bacteriophage, the phenomenon of lysogeny complicated the simple conception of phage as a virus of bacteria. Virus-free cultures of bacteria, apparently isolated from single cells, would spontaneously start producing phage that made plaques on suitable hosts. Bacterial cultures that could be grown for generations suddenly would lyse, or produce phage that could lyse other cultures. These cultures that seemed to possess the latent ability to produce phage and sometimes lyse were thus termed lysogenic.

The phenomenon of lysogeny occupies a special chapter in the early history of phage. Both the complexity and confusion it generated and the decision by Delbrück to eliminate lysogeny as a central problem for APG attention require careful examination. On one hand, the study of lysogeny was initially excluded from the research program of the APG because it seemed to complicate and compete with the consensus that the central problem of phage biology was to understand phage reproduction. On the other hand, the study of lysogeny provided deep insights into mechanisms of gene regulation and eventually provided much of the mechanistic understanding that united gene reproduction with gene function, transmission genetics with physiological genetics.

Félix d'Herelle himself observed lysogenic cultures, but he viewed them as mixed cultures of phage and bacteria in some sort of equilibrium where the phage were infrequent enough to escape detection. Others, Jules Bordet in particular, focused on lysogenic bacteria as the source of phage, as a lytic substance that was bacterial in origin and auto-inducible by the same lytic substance. This complexity was the source of much of the acrimonious debate between d'Herelle and Bordet. The nature of lysogeny was central to the developing field of bacterial genetics in the 1930s and became a key property in understanding bacterial diversity in *Salmonella* and gene regulation in *E. coli* in the 1940s and 1950s.[1]

The early studies of lysogeny were hampered by two main problems: the lack of a standard bacterial model and the lack of genetic or biochemical tools for its study. While most researchers appreciated the diverse behavior of individual phage and bacterial isolates, there was no general agreement on what might be a standard "system." This lack of agreement on methods and materials, of course, was characteristic of the pre-framework state of the field. The same for the tools of bacterial genetics such as gene mapping by recombination, stable phenotypic properties representing known genetic constitutions, and reproducible methods for induction of phage production by lysogenic cultures all were impediments to progress in study of lysogeny.

Soon after the original discovery of phage in 1917, Bordet and Mihai Ciucă (1883–1969) in 1920 found phage associated with "leukocytic extracts" (peritoneal exudates containing white blood cells induced by injection of bacteria into mice) and concluded that the bacteriophage phenomenon was the result of "vitiation" (bringing to life) of a bacterial autolytic principle. Marcel Lisbonne (1883–1946) and Louis Carrère (1892–1974), in Montpellier, noted that their strain of E. coli when mixed with several strains of Shigella would produce lysis of the Shigella strains, but not of the E. coli strains. They suggested that the E. coli strain, while not susceptible to lysis, produced an antagonistic principle that was serially transmissible and would lyse Shigella strains. They referred to the culture of E. coli as possessing the "capacity of spontaneous lysogenesis" (pouvoir lysogène spontané).[2]

The early phage papers used the term "lysogenesis" broadly to describe phenomena in which cultures unexpectedly cleared, apparently spontaneously or upon some manipulation such as mixing organisms as described by Lisbonne and Carrère. By the latter years of the 1920s, however, the E. coli strain isolated by Lisbonne and Carrère was starting to be the standard exemplar for study of this aspect of phage biology. E. coli Lisbonne, as it was called, was studied in Brussels by Bordet, in Paris by d'Herelle, and in New York by Earl B. McKinley (1894–

1938).[3] Puzzling results with these strains needed to be clarified: spontaneous and acquired resistance to phage was a constantly confronted phenomenon, both in laboratory experiments and in clinical application of phage therapy. What was the nature of this resistance? With only a rudimentary idea that bacteria were genetic organisms, researchers attributed resistance to diverse causes. Gradually, however, in the 1920s, the picture emerged that the lytic principle, the phage, was produced by the lysogenic strain that was usually resistant to the phage; another non-lysogenic, sensitive microbial strain was required to observe the action of the induced phage. The control and nature of the phage in the lysogenic strain, however, was hotly debated. D'Herelle, committed to a strict viral view of phage, argued that lysogenic strains were in some way "contaminated" and that other strains were "ultrapure," in his terminology. On the other hand, some scientists, such as Bordet, believed that the lysogenic state proved that bacteriophage was a bacterial product in the nature of a lytic enzyme.

One common goal of the few phage researchers who worked on this problem was to devise experimental strategies to sort out the possibility of ordinary contamination from some sort of phage-bacterium symbiosis. As early as 1925 Eugène Wollman distinguished two forms of phage-resistant bacteria: one form was a lysogenic strain such as that of Lisbonne and Carrère that could still produce infectious phage, and the other form was a strain that grew up from secondary cultures that survived massive phage infections. Not all phage-resistant bacteria were the same.[4]

The first clarity in the study of lysogeny came from the few workers who relied on *Bacillus* strains rather than the intestinal bacteria such as *E. coli* or *Salmonella*. The spore-forming bacilli provided a crucial test for one aspect of lysogeny: does the phage exist intact inside the bacterium in some sort of persistent or smoldering infection, or is it not present as phage at all? Thus, it was not the medical bacteriologists such as d'Herelle or Bordet who started to clarify lysogeny, but bacteriologists working with environmental or industrial organisms. Philip B. Cowles (1899–1978) at Yale and Louis Edmond den Dooren de Jong (1890–1980) in Rotterdam exploited the heat-resistance of spores of *Bacilli* to distinguish between bacteria and phage in simple experiments. Since free phages were sensitive to heating above about 70C, and yet bacterial spores could withstand 90C–100C heating, they reasoned that any adventitious free phage or fully formed intracellular, but latent, phage would be killed by heating to 90C. However, heated spores of certain *Bacilli* that were "lysogenic" gave rise, upon germination, to bacterial colonies that continued to be lysogenic—that is, they liberated phage that made plaques on indicator strains that were known to be sensitive to the particular phage strains. Somehow, the capacity to produce phage resided in the bacteria not as phage but as a heat-resistant part of the bacterial cell itself.

This elegant distinction between bacterial viability and phage viability was not possible for non-sporulating bacteria such as *E. coli* or most other enteric microbes. The use of *Bacillus megaterium* for lysogeny research became established by these early studies, and *B. megaterium* became the favorite organism of the Wollmans and later André Lwoff at the Pasteur Institute.

The Wollmans repeated and extended the approach of Cowles and den Dooren de Jong and hypothesized that the capacity to produce phage was part of the hereditary apparatus of the bacterium. While this suggestion seems eminently reasonable to the modern reader, in the 1930s bacterial heredity was something of a mystery. Bacteria, without a known mating system and devoid of a compartmentalized nucleus with visible chromosomes, were viewed as having a distinctly different sort of genetics from eukaryotic cells.[5]

In Australia, Macfarlane Burnet, a medical bacteriologist using clinical isolates, clarified lysogeny by tracing its transmission as a heredity property through various isolation procedures and suggested, in analogy to the then-current ideas from embryology, that the cell harbored the *Anlage* (German: a laying out; plan) of the virus in latent form that could be passed from cell to cell, even if no infectious virus could be detected.[6]

Interpretations such as those of Burnet and the Wollmans, however, did not go unchallenged. In addition to the Bordet school, which viewed phage as analogous to a secreted enzyme, another Nobel winner, John H. Northrop, and notably his student Albert Krueger, working at the Princeton Laboratories of the Rockefeller Institute, viewed phage in the context of Northrop's famous work on activation of proenzymes, such as the trypsinogen to trypsin, pepsinogen to pepsin, and fibrinogen to fibrin reactions. For Northrop, lysogeny was something like a slow-motion version of lytic bacteriophage production. In particular, Northrop and Krueger were not convinced by the single-step growth experiments of d'Herelle, Burnet, and Ellis and Delbrück. Working with a lysogenic strain of *B. megaterium*, Northrop demonstrated a close parallel between the continuous secretion of the extracellular enzyme gelatinase and the continuous increase in phage titers in cultures of this organism.[7]

Indeed, it is now hard to imagine the context of this early work in sociocultural terms: with two Nobelists (Bordet and Northrop) and the authors of the major textbook in opposition to the virus theory of phage, it must have been a daunting challenge for isolated, young, and unknown scientists such as Burnet, Delbrück, Luria, and a few others to soldier on.[8]

In one of his discipline-building manifestos, Max Delbrück summarized his rather ambivalent views on lysogeny in
1942:

> [T]here are bacteria which are virus carriers. Such
> strains harbor the virus and secrete it, but show no
> pathological symptoms. Such strains are called "ly-
> sogenic," because filtrates of such cultures of such
> strains will always contain the virus, as can be shown
> by the use of some lysable strain as indicator. . . .
> In lysogenic strains the secretion of the virus is anal-
> ogous to the secretion of an extracellular enzyme.
> Northrop has studied this analogy more closely with
> a strain of B. megatherium [sic], which is lysogenic
> and produces the extracellular enzyme, gelatinase.
> . . . In lysogenic strains the symbiosis between the
> virus and the carrier is a very intimate one; it is
> a coordinated growth and the host cannot by any
> means be divested of its physiological function (Bur-
> net). Every cell of the host strain is lysogenic, and
> in spore formers the symbiosis is carried through
> the spore stage (Wollman and Wollman 1939). . . .
> The phenomenon of lysogenesis has its analog in
> plant and animal viruses where it is called "indige-
> nous virus." . . . Andrewes (1939) has recently re-
> viewed the possible relevance of these findings to
> the virus aspect of the cancer problem.[9]

Only three years into his conversion to phage research,
Delbrück recognized the value of clearly stated, unambiguous
positions, yet he did not hesitate to speculate on the global
relevance of phage research to all of biology. By 1946 he was a
bit more tentative: "We are not entirely convinced that virus is

liberated in these strains from the cells in which it multiplies without destroying the host cells."[10]

Alfred Hershey, along with his mentor, the early phage researcher Jacques Bronfenbrenner, as co-author prepared a review on bacteriophages in 1948 for an influential text, *Viral and Rickettsial Infections of Man,* edited by Thomas Rivers (1888–1962), and they were vaguely dismissive of the importance of lysogeny: "It must be concluded, however, that the phenomenon of lysogenesis, frequently cited as evidence for spontaneous intracellular origin of virus, can equally well be explained as one type or another of association of exogenous virus and incompletely susceptible bacterium."[11]

In the postwar period of phage research, three researchers and their laboratories stand out for their dogged interest in lysogeny: Sir John S. K. Boyd (1891–1981) in England, Esther Lederberg in the United States, and André Lwoff in France. While Lederberg and Lwoff are well known in the phage community, Boyd is not. This is so, no doubt, because he was interested in phage as a factor in human disease, especially dysentery, rather than as biological entities in their own right.[12] His work, however, in the early 1950s was influential, especially on the Lederbergs, who saw the *Salmonella* group of organisms as a promisingly diverse group of enteric organisms with which to test the generality of results from *E. coli.*

In 1916 Boyd worked with F. W. Twort (1877–1950) in Macedonia, studying bacillary dysentery, and this contact introduced him to phage very early; later he dabbled in phage therapy. In 1950, Boyd, now working at the Wellcome Laboratories of Tropical Medicine in London, published his first paper on lysogeny in *Salmonella.* By 1956 he had become an established authority and published a thirty-six-page review devoted to lysogeny in *Biological Reviews.* In this review, he supported the view of lysogeny that had developed in the previous decade:

"A *lysogenic bacterium* is one which possesses and transmits to its progeny the power to produce phage. It does so by virtue of the *prophage* which it contains." Boyd argued in this review that "Much of this work [leading to understanding phages] has been carried out on the T series of coliphages, a group of virulent phages collected from various sources by Demerec & Fano (1945), which by common consent has been adopted for study by most of the American workers. More recently, increasing attention has been directed to the lysogenic bacterium and to the problems it raises, and, as a consequence, a more balanced picture of the cycle of development—or life cycle—of the bacterial virus is beginning to emerge."[13]

How did lysogeny evolve from Delbrück's 1942 enzyme secretion analogy to the "prophage" model less than ten years later? Although the Wollmans, Burnet, den Dooren de Jong, and Cowles, among others, established the fact that every cell in a lysogenic culture carried the lysogenic property, and, indeed, it could survive sporulation and germination in spore-forming strains and thus has the characteristics of a heritable property, the nature of this property, Burnet's *Anlage*, was unclear.

One impediment to clarity was the tendency of bacteriologists at the time to think in terms of bacterial cultures rather than bacterial cells. The behavior, whether it was growth, production of phage, or lysis, was seen in terms of the macroscopic properties of the entire culture. While no one doubted that the culture was composed of a large population of individual bacteria, rarely was attention directed to the individual cell and its properties and behavior.[14] In the 1930s this viewpoint began to change, and this shift was driven, in part, by the developing understanding of bacteria as organisms with comprehensible genetic behaviors. This interest in the cell as the unit of analysis spread into phage biology, for example, in the idea of the plaque assay, the notion of an "infective center," and the single-burst experiment. All these experimental approaches pointed to the need to study the behavior of individual cells rather than the aggregate behavior of the culture.

Two crucial kinds of experimental evidence led to this evolution in thinking about lysogeny in the late 1940s and early 1950s. The first approach was taken by Lwoff and his colleagues in Paris. Lwoff, a microbial physiologist and an early commentator on the Wollmans's work on lysogeny at the Pasteur Institute, devised an experimental approach to the problem of phage production by lysogens. After World War II, during which the Wollmans perished in Auschwitz, Lwoff took up the study of phages at the urging of Delbrück.

Lwoff's interest focused on microbial cells, and less on the mysteries of the phage itself. The interplay of bacterial growth and phage production seemed to appeal to him as a problem in microbial growth, not virology. In a series of simple experiments, focused on single cells, Lwoff and Antoinette M. Gutmann (1925–2016) swept away decades of vague conjectures and uncertainty. They proposed a clear conception of lysogeny that illuminated the understanding of cultural behavior in terms of the collective behavior of individual cells. They also reformulated and forcefully stated a concept of lysogeny that incorporated the *Anlage* of Burnet and the hereditarian notions of the Wollmans. They also refuted the secretion hypothesis of Northrop and others. In a key publication in 1950 they summarized their experiments and argued that each cell in a lysogenic culture harbored a phage precursor of some sort, which under certain conditions would be activated to produce phage with concomitant cell lysis.[15]

In a series of experiments with single cells of *B. megaterium* in small droplets observed under a microscope, Lwoff and Gutmann distinguished between phage released by continuous secretion from intact cells and phage released by cell lysis, thus settling the long-standing issue raised by Northrop and others. By direct observation and micromanipulation of cells and assays of tiny samples of the droplets, the behavior of the individual cell was correlated with phage appearance. In an impressive number of trials, phage only appeared when a cell lysed and disintegrated. In addition, the same micromanipulation approach allowed them to transfer one of the two new daughter cells to fresh medium and at the same time assay the other cell for phage production. Even after nineteen phage-free transfers, all daughter cells still retained the lysogenic property, leading to the conclusion that the phage is propagated in concert with the cells, intrabacterially. For the first time, a description of the lysogenic state was formulated that was both clear and yet agnostic about the nature of the newly christened prophage (protobacteriophage). This model was heuristic but also couched in terms familiar to the majority of phage biologists. It fit the evolving framework of phage and did much to make lysogeny respectable (figure 8).

Although the studies of Lwoff and Gutmann established useful new facts about lysogeny because they dealt with cells and phages as the unit of analysis, they did little to speak to the intracellular details of lysogeny. A different approach, pioneered first by Esther Lederberg and shortly thereafter by Giuseppe Bertani (1923–2015), was to combine genetic manipulation of the host strains with phage biology.

In 1949, Esther Lederberg isolated a mutant of *E. coli* K12 that was a sensitive indicator strain for phage produced by her wild-type *E. coli* K12, which showed that the wild-type was an unsuspected lysogen. Since the Lederbergs were focused on the study of the genetics of bacteria, and *E. coli* K12 in particular, the property of lysogenicity became, for them, a genetic

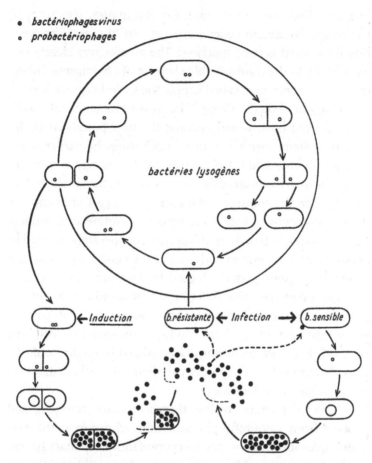

- bactériophages virus
- probactériophages

bactéries lysogènes

← Induction   b.résistante ← Infection → b.sensible

Figure 8. "Diagram showing the evolution of bacteriophage and pro-bacteriophage in susceptible and lysogenic *Bacillus megatherium*" (Lysogeny scheme from André Lwoff and Antoinette Gutmann, "Recherches sur un *Bacillus megaterium* lysogène," *Annales de l'Institut Pasteur* 78, no. 6 [1950]: 736) property. Did lysogenicity, and, for that matter, sensitivity to the phage, behave in genetic crosses just like more conventional bacterial genetic markers? The answer was clearly yes. By analogy to the transmissible element of cytoplasmic inheritance in *Paramecium*, called kappa, the Lederbergs named the new phage lambda (λ). Phage λ, because of its genetically well-characterized host, would become the new paradigmatic lysogenic system, nearly completely eclipsing the earlier study of the Lisbonne and Carrère strain of *E. coli* and *B. megaterium*. Soon on the heels of the discovery of λ, Bertani, who had stumbled on the confusion of lysogeny in 1949, found another lysogenic system in enteric bacteria with interesting properties not shared with λ. Bertani's system, deliberately selected to avoid direct competition with that of the Lederbergs, used the original lysogenic system studied by Lisbonne and Carrère, a temperate phage—that is, a phage that could infect, but not lyse, a bacterium and convert it to a lysogen, P2, and *E. coli* C as the host bacterium. These two lysogenic systems, with host bacteria that were becoming standardized biological objects, provided tools to study lysogeny in detailed mechanistic ways not possible in earlier times.[16]

Over the course of less than five years, lysogeny had evolved from a doubtful phenomenon, whose existence was widely questioned, to an accepted part of the disciplinary framework of the young APG. This was made possible by the reframing of lysogeny in new terms and concepts already vetted and accepted by the phage community.

As Richard Burian and Jean Gayon, as well as Nadine Payrieras and Michel Morange, have convincingly illustrated, the particular focus at the Pasteur Institute provided a fertile environment for the integration of the concepts of cellular and viral physiological genetics, exemplified by the subsequent work of François Jacob (1920–2013), Jacques Monod (1910–1976), and

Lwoff.[17] In the United States, by contrast, the focus on phage, its chemistry, and its intracellular structure and replication complemented the French work, thus illuminating the molecular nature of the prophage and its relationships to the host genetic structures. The acceptance of lysogeny as a legitimate topic in phage research represented an evolution—perhaps one might say a maturation—in the framework of the APG. It also provided an important bridge between the American and French phage workers, allowing each group to retain certain basic viewpoints while at the same time reframing phage biology in new ways. These new directions allowed the APG to overlap with mainstream microbial genetics in a manner not originally envisioned by the physicist-founders of the APG. This was especially true in terms of allowing biochemists to begin to dissect the inner workings of the "black box" that was the physicists' view of the bacterial cell.

The study of lysogeny, especially phage λ, would yield crucial insights for cell biology and virology in particular, because of the importance of so-called lateral gene transfer, in which genes are transferred between cells without cell division or conventional mating behaviors. Pieces of DNA, usually circular molecules, once inside cells, can be inserted into the cellular chromosome by recombination mechanisms that were first discovered for phage λ, in work to explain the way that phage makes a bacterium into a lysogen. This mechanism of integration by a single reciprocal recombination event between the circular phage (or other DNA) molecule and the resident chromosome was proposed in 1963 by Allan Campbell to explain why the genetic maps of the vegetative (lytic) form of λ differs (it is inverted) from the genetic map of the same phage when it is in the latent (chromosomal) form in the lysogen. The "Campbell model" (figure 9) for integration led to understanding how antibiotic resistance genes are spread in bacterial

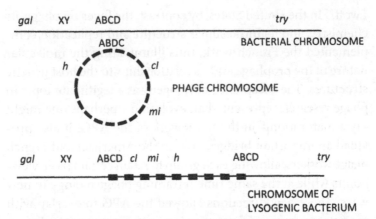

Figure 9. The Campbell model for temperate phage integration into the bacterial genome. Reciprocal recombination between the ABCD site on the bacterial chromosome and the ABCD site on the circularized phage chromosome results in the linear integration of the (inverted) phage chromosome into the bacterial chromosome resulting in the lysogenic state.

populations and how retroviruses, including HIV, integrate into eucaryotic cells.[18]

Lysogeny had another salient effect on molecular genetics when it was discovered that bacterial genes could be transferred between bacteria by phages. When Norton Zinder (1928–2012) was attempting to repeat Lederberg's bacterial mating experiments from *E. coli* with the related bacterium, *Salmonella sp.,* the gene transfers did not require the bacteria to be physically mixed, but rather the bacterial genes were transferred by a "filterable agent" in the culture fluid of the donor strain. Zinder and Lederberg concluded that the bacterial genes were being transferred by a phage-like particle produced by the donor strain, and they called this process "transduction" by analogy with the classic case of DNA-mediated gene transfer, called transformation by Avery, McCarty, and McLeod.

It appeared that some of the temperate phages induced from
the lysogenic bacteria had incorporated pieces of host cell DNA
into phage, or phage-like, particles, which was then transferred
to susceptible recipient cells where these foreign genes could
be incorporated, presumably by some sort of genetic recombination process. Because only relatively small fragments of
the bacterial genome can be carried between bacteria this way,
only closely linked genes can be co-transferred. Zinder and
Lederberg's P22 *Salmonella* phage transduction system was
soon exploited by Milislav Demerec and his colleagues in fine-
structure gene mapping studies in *Salmonella*.[19]

Soon several variations on this theme were discovered.
Phage-mediated gene transfers of several types were found. In
contrast to P22, which only carried small fragments of bacterial DNA, phage P1 of *E. coli* apparently assembled in such a
way that phage heads could be completely filled with bacterial
DNA and no phage DNA, using the phage structure to deliver
only bacterial genes. P1 phage could thus transfer larger fragments and expand the reach of co-transduction experiments
to map more distantly related genes on the bacterial chromosome. Other phages, including λ, were found to be restricted
to transduction of a single, specific host gene. M. Lee Morse
found that λ phage stocks contained a very minor population
that could transduce the host gene for galactose metabolism,
and additional studies suggested that the host DNA was carried
at the expense of some dispensable phage genes.[20] This transducing phage was called λdgal for "defective, gal-containing,"
sometimes abbreviated λdg. This type of gene-restricted transduction was later understood in terms of the Campbell model
for prophage integration where the phage integration site on the
host DNA was limited to a region near the galactose genes, and
hence the phage DNA excision step, when imprecise, allowed
for only the nearby gal genes to be picked up for transduction.

Transduction of bacterial genes by temperate phages is a widespread phenomenon, and it is the basis for much of the lateral gene transfer between compatible bacteria, allowing for evolutionary gene flow that complements classical reproductive gene transmission.[21] The use of virus particles to package and deliver non-viral genes is the basis of much current gene therapy as well as new vaccine technologies.

Of course, the major challenge that lysogeny presented to the APG was the problem of what kept the prophage in its latent form, inhibiting phage reproduction, cell lysis, and release of progeny phage. Lytic phage were simple parasites: they attacked their hosts, took them over, consumed their resources, and produced more phage. Temperate phages, on the other hand, became quiescent, but they were potentially able to become virulent and reproduce and lyse the host. What kept them quiescent and how did they become active? It was the study of this aspect of phage biology that would link genetics and reproduction to gene expression and cell metabolism, transmission genetics with physiological genetics, and the American phage school with the French phage school.

In the early 1950s, soon after the discovery of phage λ, several phage workers discovered plaque morphology mutations of λ and other temperate phages that produced clear plaques instead of the usual turbid plaques.[22] The clear plaque mutants were unable to form lysogenic bacterial strains, thus mimicking the classical lytic behavior of the T-phages. Not only were these mutants useful in genetic mapping of these temperate phages, but they became key to the puzzle of how the lysogenic state was maintained in host bacterium. One puzzle in lysogeny was the phenomenon of "immunity"—not classical immunity of higher organisms, but simply resistance of lysogenic strains to infection by the same phage that the lysogen could produce when induced to do so. When it was found that lysogens were not immune to the clear plaque mutants, a mechanism of immunity was proposed that involved some factor in the lysogen that could inhibit the superinfecting phage. If the superinfecting phage was a clear plaque mutant, able to grow on the lysogen, it must lack a functional target for the inhibiting factor. This mechanism immediately suggested a way for the prophage to be held in check inside the lysogenic bacterium: the inhibitory factor acted on the prophage just as it did on the superinfecting added phage. What started as an experiment in gene mapping had become an experiment in cell physiology. These experiments were taking place at the very same time and, indeed, sometimes in the very same laboratory (the Pasteur Institute in Paris) where the phenomenon of adaptive enzyme synthesis was being elucidated and the repressor-operator (operon) hypothesis had just been elaborated.[23] With further study, it became clear that the control of lysogeny was a nearly exact replica of the cellular machinery by which the many adaptive gene expression systems were regulated in bacteria. In this way, phage λ and the problem of lysogeny would usher in a grand synthesis of the two contesting themes in twentieth-century genetics, the American school of Morganic heredity and the European school of Roux-Driesch embryology. Genes do both.

10

The Challenge of Phage Diversity

The "Phage Treaty" of 1944 was intended to focus phage research on a small set of standard phages, designated T1–T7, and a single host bacterium, *E. coli* strain B, for three reasons: to allow comparability between laboratories, to facilitate the goal of complete understanding of at least a few phages, and to enforce community membership based on sharing knowledge and materials, for example, strains and mutants, among the APG. This practice worked well at the beginning, but as the field grew both in size and in understanding of phage biology, the list of acceptable phages as standard increased.

The first new member of the phage list was the temperate phage λ as just discussed. The phenomenon of lysogeny could not be studied with the strictly virulent T-phages. The expansion of the list of acceptable phages did not challenge the basic framework of shared commitments of the APG but did require an evolution in the technology to be part of the revised shared commitments. In other words, although the shared commitments of the APG included the belief in the need for an agreed-upon set of phages to be investigated, along with certain norms for the selection and deployment of the particular phages, the shared framework did not include, as fundamental, any particular phage or set of phages. As the research program matured and new problems arose, different phages might have to be examined that were most likely to provide the needed technology.

One of the first of the second generation of phages to be legitimized by the APG was a phage designated φX174. By the 1950s, several lines of research had firmly established that the phage particle sizes and the DNA contents of the T-phages were relatively large (by virus standards). Delbrück and Robert L. Sinsheimer (1920–2017) decided, in about May 1953 according to Sinsheimer's recollection, that a much smaller, simpler phage would be useful to study. This new interest in phage DNA belied the growing incursion of biochemistry into the core thinking of the APG. A survey of known phages, for which some size information was known, yielded two candidates: a phage (φX174) isolated on an *E. coli* host by Vladimir Sertić and Nicolai Boulgakov in d'Herelle's laboratory in 1935, and another phage (S13) isolated on a *Salmonella* host by Burnet.

When Sinsheimer, a biophysicist, obtained samples of both phages, he found that φX174 was easier to grow in large quantities, so he concentrated on that phage. In his initial characterization of this phage, its DNA behaved anomalously in several ways: its sedimentation rate conflicted with other measurements of its molecular weight, and its optical properties upon heating did not behave as was typical for normal, double-stranded DNA. In addition, the nucleotide base composition did not conform to the base-pair ratios of the normal Watson-Crick pairings based on Chargaff's Rules. As Sinsheimer noted, it was an "inescapable conclusion that a single virus particle carrying a single-stranded DNA molecule can initiate infection." Other physical studies, coupled with chemical studies of φX174 DNA, yielded the unexpected finding that the DNA was not only single-stranded but also in the form of an endless ring. Small, single-stranded, circular DNA: clearly a novel addition to the collection of approved phages for the APG.[1]

In the early 1960s, the unification of transmission genetics and physiological genetics was in full swing, and phage biologists had started to think about the way phages grew and developed, no longer simply preoccupied with the problem of faithful gene reproduction. Phages with various gene systems would allow study of gene expression and regulation both in vivo and in vitro. Phages with small DNA and (soon to be discovered) RNA genomes provided sources of relatively small homogeneous molecules for biochemical experimentation. Study of the reproduction (more frequently called replication, by then) of phages with double-stranded DNA in comparison with the single-strand DNA phages would lead to better understanding of the way genes were copied from generation to generation. As Sinsheimer recalled in 1966, "More and more, the participation of ring structures in DNA replication—cellular or viral—is seen to be widespread. . . . And here study of the reproduction of φx DNA rings may serve as a miniature, more readily analysable model for DNA ring replication in general, an amenable experimental system with which to explore many still mysterious facets of DNA synthesis."[2]

Other small DNA-containing phages were soon isolated and studied for comparative reasons. Burnet's phage S13 was favored by some, notably, Irwin Tessman (1928–2016) and Ethyl Tessman (1928–1986) and their colleagues, whose work with this phage complemented that of Sinsheimer and his followers. Both groups were adopted into the APG as φX174 became legitimized as an object worthy of study. Soon other such small DNA phages were recognized.[3] G. Nigel Godson (b. 1936) isolated a series of phages selected by small size and ability to grow on *E. coli* C as does φX174, one of which (G4) proved especially useful for in vitro DNA synthesis studies.[4]

A different group of DNA-containing phages was discovered when Tim Loeb (1935–2016) and Norton Zinder were seeking phages that had a specific host range for *E. coli* strains that carried the fertility factor, F, often called "male" *E. coli*, based on their ability to donate their DNA to other, recipient "female" strains. They found phage isolates that would make plaques on male bacteria but not on female strains. Two of their isolates turned out to have previously unknown characteristics. The structure of one of these, f1, was found to have a DNA genome, again, single-stranded and circular, similar to φX174, but it was filamentous or rod-shaped, quite different from the spherical shapes of φX174 and the T-phages (with the addition of tails on several of the T-phages). Soon other groups found that some of their male-specific phage isolates were similar to f1. These related phages, for example, fd and M13, became widely used by certain research groups, not closely connected to the APG.[5]

One isolate, designated f2, had an RNA genome. While tobacco mosaic virus and several animal viruses such as polioviruses were known to have RNA genomes, f2 was the first bacteriophage found to have an RNA genome.[6] The conjunction of this discovery of an RNA phage and the discovery of the role of RNA as a messenger molecule in gene expression made RNA phage an immediate tool in research on aspects of protein synthesis. Clearly, RNA phages were candidates for the expanding list of APG-recognized phages.

The approach used by Loeb and Zinder to isolate phages that might contain RNA was quickly adopted by others, and soon several other groups found similar male-specific, RNA-containing phages. R17 phage was isolated by William Paranchych (1933–1995) and Angus F. Graham (1916–2008), and

MS2 was found by A. John Clark (b. 1933).[7] One might expect, based on the culture of the APG, that one (or a few) of these new RNA phage isolates might become "standardized" as was the case for φX174, but because of the outsider status of several of the RNA phage workers and because the APG norms for material and information sharing were not fully recognized, an accepted isolate, representative of the RNA phages, did not emerge.[8] As a result, all three phages, f2, R17, and MS2, had their followers.

Comparative studies, of course, have been a mainstay of biology for centuries, and constraints and limitation on diversity have been debated in comparative biology at the same time. It was no different for the APG. Since the three widely studied RNA phages, f2, R17, and MS2, were all related, at least serologically, the discovery of an unrelated RNA phage in 1961 by Itaru Watanabe (1916–2007), designated Qβ, provided a useful comparative phage; it was rapidly exploited in biochemical studies of RNA replication as well as experimental phage evolution.[9] Fortunately, these phages are sufficiently similar, and the questions they were used to investigate were similar enough, to make this diversity tolerable, perhaps even appreciated.

The growing diversity with respect to experimental material that was accepted by the APG was an indication of increasing scientific maturity, and it was a forerunner of the evolution of phage biology into a foundation of molecular biology. At the beginning, the Phage Treaty was essential to impose some order on the individuality of each investigator so that, in the fact-collecting phase of the research program, the various results could contribute to a common, universally relevant body of knowledge. Later, this knowledge became sufficiently detailed and robust that researchers were confident in extending and extrapolating it to new phage types that had useful, and perhaps novel, specific properties. This incorporation of diversity did not threaten the field with the earlier chaos because accumulated knowledge allowed exceptions to be accommodated without challenges to the overall well-confirmed theoretical edifice.

11

The French Connection

Centrifugal Impulses and Expanding Influences

lthough the discovery and first extensive investigations of bacteriophages took place in prewar France, the subsequent developments in phage research were centered in North America. There remained, however, scattered interest in phages both in Europe and in parts of the Soviet Union and Australia. Some of this phage work became closely allied with the APG; however, there were several significant instances of more independent research programs. In this chapter we will consider some of these linkages and reciprocal influences.

As Donald Fleming and Bernard Bailyn noted in their seminal work on émigré scientists in the Americas starting just before World War II, European scientists changed the landscape of American science, and in no place was it so apparent as in the founding and development of the APG.[1] Max Delbrück and Salvador Luria both arrived in the United States just before the U.S. entry into World War II, and both settled in their new country as permanent residents, Delbrück as an intellectual dissenter from Nazi Germany, and Luria as a Jewish refugee from Fascist Italy. The war years saw few, if any, potential phage workers arrive in the United States, but immediately after the war there was a steady stream of European visitors who came both for personal reasons and as emissaries sent by their governments to catch up on wartime scientific advances and to bring this new knowledge back home. It is important to distinguish this group of émigré phage researchers who left Europe and established themselves in the United States from those who are the subjects of this chapter—Europeans who established collegial and collaborative connections with the APG.

In the immediate postwar period there were important connections forged between the fledgling APG and the developing school of French molecular biologists. Although the Pasteur Institute was a center of French bacteriology, it had always had a medical orientation, and this was especially true of the virology section under the leadership of Pierre Lépine (1901–1989) in the years just before and after World War II. Medical bacteriology and vaccine development and production were the main foci for postwar development. Lépine saw certain instruments and technologies, especially the electron microscope, as crucial to the important medical tasks of diagnosis and identification of viruses. Others, such as André Lwoff and his young colleague Jacques Monod, argued for new directions in virology, toward basic biological understanding, a "demedicalization" as the historian Jean-Paul Gaudillière has called it.[2]

In 1946 Monod attended the Cold Spring Harbor Symposium on Quantitative Biology that was devoted to microbial genetics, and he came back impressed by the fact that the American research had very little relation to medicine, per se, and that genetics in the Morgan sense was dominant. He and Lwoff, after their unsuccessful attempts to reorient the Pasteur Institute in less medical directions, turned to the Americans for support and legitimacy that they could not achieve in France. As the historian Doris Zallen has written, the relationship between Louis Rapkine (1904–1948) as a respected voice for French science and the Rockefeller Foundation in the United States would become crucial in the postwar reconstruction of French science.[3] Rapkine became a key connection between the APG and the French biologists, but he was almost unknown to most of the Americans because his role was political rather than scientific. A Russian Jewish refugee who spent his youth in Canada, Rapkine relocated in the mid-1920s to Paris, where he worked as a biochemist in the College du France on problems related to sulfur metabolism. His clandestine political activities from 1936 onward assisted European refugee scientists, and from 1940 to 1944 he worked in the United States to support the Free French Government-in-Exile and with the Rockefeller Foundation to rescue numerous French scientists who then resettled in North America. After World War II, Rapkine returned to Paris and established the department of cell chemistry at the Pasteur Institute, where he was a trusted liaison to the Rockefeller Foundation. Almost immediately after World War II, the Rockefeller Foundation awarded $100,000 to the French CNRS (*Centre national de la recherche scientifique*) to support foreign visitors' participation in French scientific conferences. Among the first to take advantage of this opportunity were André Lwoff and Boris Ephrussi (1901–1979), who organized a meeting titled "Biological Entities Endowed with Genetic Continuity" in June 1948.[4] Among the attendees at this meeting were the phage biologists Max Delbrück and Harriet Taylor.

Another Pastorian, François Jacob, later recalled:

At the time, we often went to the United States. As
a matter of fact, I think that after the war Lwoff and
Monod were the first ones to rush to the United
States. I think they had both received Rockefeller
grants. So, there were links. They also got money
for the lab from Rockefeller. I also got some [money]
straight away, as soon as I started to emerge and
write papers. So, it worked out very well. [Inter-
viewer: And thankfully, because you weren't getting much help in France?] Not much, the Pasteur
Institute didn't get any, and from the extra-Pasteur
Institute, from the ministries we got very little. So,
we had a little American money which was valuable.
Which was especially valuable because we bought a
lot of things in the United States. Machines.[5]

Two young French scientists came to the United States
for extended research visits and became pioneers of this connection. The first was Raymond Latarjet and the second was
Elie Wollman. Both became leaders in French molecular biology. Both had strong French scientific pedigrees as the offspring of well-known French biological scientists—Latarjet, the
son of the famous French surgeon and anatomist André Latarjet (1877–1947), and Wollman, the son of two Pastorian bacteriophage microbiologists, Eugène and Elisabeth Wollman, both
victims of the Holocaust. Wollman was also the protégé of the
more senior French microbiologist André Lwoff, who was a
third key connection between France and North America.[6]

Raymond Latarjet as a young physician had taken an
interest in radiation biology just before the war. For a short
period, Latarjet worked in the Radium Institute at the same
time as Salvador Luria did, before Luria emigrated to the United
States in 1940. Immediately after the war, the French govern-

ment sent a small group of young physicians, including Latarjet, to the United States to learn what advances in medicine had
occurred during the hiatus in communication caused by the
Nazi occupation. Latarjet was the only non-clinician, and he
was assigned to Columbia University, where he found no opportunities for research in his field of radiation biology. There
he learned that Luria, formerly a guest researcher at Columbia but at that time a young faculty member at Indiana University, spent his summers at Cold Spring Harbor Laboratory
on nearby Long Island. Based on his casual acquaintance with
Luria in Paris, Latarjet approached him and proposed a collaboration based on work that Latarjet had already started back
in France. These experiments involved using target theory approaches to count the number of intracellular infective phage
during the phage growth cycle. Since a phage-infected bacterium, when plated on a lawn of sensitive bacteria, gave only one
plaque, no matter how many infectious phages were inside the
single infected bacteria, he reasoned that this plaque-forming
ability of a bacterium with many phage inside would be harder
to inactivate by x-rays than an infected bacterium with a single
or a very few infectious phages inside. By measuring the target "hit number" from survival curves for this plaque-forming
ability, assayed at increasing times after infection of a bacterium, the intracellular increase in infectious phages could be
determined. This target theory approach to study intracellular
phage growth was successful and became a standard method
prior to the widespread application of radioisotopic tracers,
which were unavailable at the time.[7]

Although Latarjet had conceived of the project and had
collected significant preliminary data, Luria convinced him
that their publication should have Luria's name first because
it would be more beneficial to him, still being in a precarious
academic position, than to Latarjet, who had a secure post

awaiting his return to Paris. Latarjet relented, reluctantly it seems, and the method became widely known as the "Luria-Latarjet experiment," often familiarly termed the "L-L experiment." Even many years later, toward the end of his life, this authorship question still seemed to pique Latarjet's ire.[8]

Latarjet's contacts with the APG continued, to a large extent, because of the central role that the target theory and radiation biology had in early phage research. As described earlier, radiation was a key conceptual and methodological link for the growing number of physicists who saw biology as a new frontier for their research.

Microbiologists at the Pasteur Institute had developed an interest in the bacterial phenomenon known as adaptive enzyme synthesis, where it was found that new metabolic enzymes were produced by bacteria in response to the availability of new nutrient substances in the culture medium. Some saw adaptive enzyme synthesis and lysogeny as related phenomena, but not, at least initially, central to the APG program. Wollman's attention was drawn to this problem, perhaps because it had been a main interest of his parents. However, most significant among the Pastorians to take up phage work was André Lwoff, a protozoologist turned bacteriologist, who clarified some of the most uncertain aspects of lysogeny as described earlier. Lwoff was born in France in 1902 to parents who had fled czarist Russia, and he became a committed internationalist. He joined the Pasteur Institute at age 19 and never left, eventually becoming a department chief. He received early support from the Rockefeller Foundation and spent time in Germany and England between the wars. His English was said to be impeccable and his writing was clear and often droll. In his recollections for *Phage and the Origins of Molecular Biology*, for example, Lwoff recounted his early discussions with Jacques Monod about metabolic adaptation: "'What could it mean?' he asked. I said this could have something to do with adaptive enzymes. The answer I received takes now full fragrance in the light of Jacques Monod's achievements. The answer was: 'adaptive enzymes, what is that?' Of course he said this in French, because we used to speak French."[9]

Elie Wollman arrived at Caltech in 1948 as a visiting scientist working with Delbrück; he would go on to provide a strong two-way channel between phage research at the Pasteur Institute and the APG. Wollman remarked in 1966:

> When I joined André Lwoff's laboratory at
> the end of the War, the advances that had taken
> place in the United States during the War were only
> just becoming known in France. Upon first reading
> the papers of the new American Phage Group, I
> found myself admiring their revolutionary prog-
> ress, at the same time, [I was] surprised by the care-
> lessness with which they treated historical matters.
> . . . In the summer of 1949, nearly the whole Phage
> Group had gathered at Caltech for several days in
> Delbrück's laboratory in order to try to fit the data
> available at the time into a general scheme. This re-
> union was the source of the "Syllabus on proce-
> dures, facts and interpretations in phage" included
> in the proceedings of the small, but pregnant, "Vi-
> ruses 1950" conference held in Pasadena the next
> March (Delbrück, 1950).
>
> One of the categories of the Syllabus assigned
> to me was lysogeny, a chapter of bacteriophage re-
> search which had been neglected (and even pro-
> nounced non-existent) by the Phage Group, but one
> on which my parents had worked for many years.
> As it happened, André Lwoff, in one of his several

avatars, had taken over the problem of lysogeny in the meantime[;] on my return to Paris in the fall of 1950, I found our laboratory greatly animated by Lwoff's new-found insights. After completing some work undertaken at Caltech with Gunther Stent on adsorption and adsorption co-factors of phage T4, I finally began, as I had intended all the while, the study of lysogenic bacteria.[10]

A special affinity soon developed between Delbrück at Caltech and phage workers at the Pasteur Institute in Paris. Perhaps this connection was facilitated by Luria's brief sojourn in Paris before emigrating to America, but more likely the empathy stemmed from the scientific culture at the Pasteur, which was, in several ways, more akin to that in the United States than to traditional French academic research. Lwoff, as Michel Morange has described him, was a "heterodox spirit" and, as a committed Darwinian, was rather isolated from the mainstream, neo-Lamarckian biologists in France. Further, the Pasteur Institute had a distinct organization and position in French culture, allowing for the development of more diverse approaches, more "heterodox" interests, and fewer (or at least different) bureaucratic pressures. Lwoff and several of his protégés seemed to find the American research "style" to their liking. One is tempted to compare the role of the Pasteur Institute in relation to North America in the twentieth century to the important influence on American clinical medicine wielded by the Paris Clinic of Pierre Louis (1787–1872) in the nineteenth century.[11]

Soon it became almost de rigueur for American phage biologists to spend a post-doctoral or sabbatical period at the Pasteur Institute. Gaudillière notes that the very first wave of postwar American visitors to Paris, such as Melvin Cohn (1922–

2018) and Arthur Pardee (1921–2019), were biochemists, probably attracted to France because of the traditional strong reputation of French biochemistry going back to Claude Bernard
(1813–1878) and even Antoine Lavoisier (1743–1794); soon, however, the microbial geneticists arrived, some of whom were
steeped in phage biology and were members of the APG. The
list of these American visitors included many of the key contributors to the community that made up the APG—phage
biologists such as Seymour Benzer, Cyrus Levinthal, Aaron
Novick, Thomas Anderson, Louis Siminovitch (1920–2021),
Margaret Lieb (1923–2018), and Charles Radding (1930–2020).
Lieb captured the essence of the French connection that existed during her stay in Paris in 1953–1954: "There were four
other Americans there on fellowships, and not much room in
the Service [the microbiology group led by Lwoff] which was
the French branch of the exclusive but unofficial group that its
members called the 'church' [a reference to the close-knit followers of Delbrück]."[12]

Traditionally, the scientific relationship between the Pasteur Institute in Paris and its offspring, the Pasteur Institute
in Brussels, has been close. The first serious recognition of
d'Herelle's discovery of phage was that of Jules Bordet, director
of the Pasteur Institute in Brussels, a recognition kept alive for
decades by the controversy fomented by André Gratia, Bordet's
protégé and heir in Brussels. It's not surprising, then, to find
that a special connection existed between Paris and Brussels in
phage research as it was developing after World War II. Gratia
and his protégé, Pierre Fredericq (1913–1984), discovered the
mysterious phenomenon of bacteriocines (they were named
for the specific kind of bacteria that produced them, for example, colicines, pyocines, and so forth)–substances produced
by bacteria that kill other bacteria of the same strain but that
are inert against the producing bacteria. These killer substances

seemed related in some way to phage, appeared promising as therapeutic agents, and presented a biological puzzle. Colicine research was a cousin to phage research, and these two branches of bacteriology developed a tenuous relationship, especially along the Paris-Brussels axis.[13]

As a rather obscure side note, it turned out that several of the young bacteriologists who ventured into phage biology in the earlier interwar period and who were early advocates of d'Herelle's conception of bacteriophages also spent time in Brussels with Bordet. Earl B. McKinley, Frederick P. Gay (1874–1939), and Bernice Rhodes were some of this group who, although they worked with Bordet and Gratia in Belgium, nonetheless adopted d'Herelle's notion of phage as virus rather than Bordet's.

Although travel across the Atlantic was expensive and not easy, it is remarkable just how much personal contact existed between the APG and phage labs in Europe. These contacts flourished in large part because of the Rockefeller Foundation's support. Warren Weaver and others were managing a top-down program of patronage to promote their vision of molecular biology by providing expensive instrumentation (such as ultracentrifuges, electrophoresis and diffusion equipment) and salary support, often conditioned on "arranged" collaborations, selected personnel, and a vertical integration of what was becoming "big science."[14]

In the 1950s, as a parallel development to the Delbrück phage course and the Cold Spring Harbor Laboratory phage meetings, Lwoff managed to organize two major phage meetings in France (1952 and 1958), both held at Abbaye de Royaumont, an elegant estate, formerly a Cistercian monastery outside of Paris, now owned by an ancient French banking and industrial family. These meetings, formally identified as the International Phage Symposia, became known informally as the

European Phage Meetings, integral in tradition with the Cold Spring Harbor Laboratory phage meetings.[15] These meetings attracted a strong contingent of North American phage workers and served to cement the community of phage workers on both sides of the North Atlantic. The French were not quite part of the APG, but nearly so, adopting similar shared research commitments.

The First International Phage Symposium was held at Royaumont from 26 July to 1 August 1952. The symposium signified the maturation of the French-American domination of phage research, but it also marked a time when understanding of phage was entering a new, more secure phase.[16] The meeting was underwritten by the National Foundation for Infantile Paralysis (in the United States), and the participants were mostly from North America and France. Max Delbrück prepared a meeting summary for the sponsors, laying out the state of the field in terms of the international research community and giving a succinct overview of several key research problems that had become clear at that point. It is useful to examine this snapshot of phage at mid-century in some detail since it provides a view of transatlantic phage relations as well as a picture of the "normal science" being pursued by the APG and their Francophone colleagues.

Delbrück summarized this meeting as follows:

> The meeting was attended by about 50 scientists from Western Europe, the United States and Canada. Most of those who attended were persons actively engaged in phage research. There were fifteen phage workers from the United States: [Mark] Adams, T.F. [Tom] Anderson, [Seymour] Benzer, [Giuseppe] Bertani, [Max] Delbrück, [Earl] Evans, [Walther] Goebel, [Alfred] Hershey, [Cyrus] Levin-

thal, [Arthur] Pardee, [Gunther] Stent, [Niccolò] Visconti, [James] Watson, [Lawrence] Weed, and [Jean] Weigle. From the Institut Pasteur in Paris came [Kenneth] Huybers, [François] Jacob, [André] Lwoff, [Pierre] Nicolle, [Louis] Siminovich, [Robert] Wahl, [Elie] Wollman and several Americans now working in Dr. Lwoff's laboratory; from Belgium, [Jacques] Beumer and [Pierre] Fredericq; from Germany, [Carsten] Bresch and [Wolf] Weidel; from England, [E.S. (Ephraim)] Anderson (Felix's lab.), [John O'Hara] Tobin, and [John] J.D. Smith (Molteno); from Switzerland, [Valentin] Bonifas and [Eduard] Kellenberger; from Italy, [Luigi] Silvestri; from Holland, [Klaas] Winkler; from Denmark, [Nils] Jerne; from Canada [Angus] Graham and [Fred] Heagy.[17]

Among those from related fields who attended the meeting were [Luigi] Cavalli, Sterling Emerson, [Boris] Ephrussi and Harriet Taylor Ephrussi, [Roland] Hotchkiss, [Jacques] Monod, and [Roger] Stanier. [Linus] Pauling visited one day.

There were no representatives from any Latin-American country because there is nobody there in these countries representing this field. There were none from the USSR and satellites. From what we know of phage research in these countries, this lack of representation was not our loss. There were none from Japan, where this field is just starting to receive attention. . . .

Delbrück identified three themes of the meeting: 1. "The three states of phage," 2. "The transitions between the Three States," and 3. "From prophage to vegetative phage." These themes seem to form a

research program for the APG if not for the entire
community of phage researchers. He also mentioned several subthemes including "The types of
phage: Temperate and Virulent," "Structure of viruses," "Phage receptors on the bacterial surface,"
and "Source and Fate of Virus Material."

Delbrück summarized clearly and succinctly
the concepts involved in understanding the relationships between the phage and its host. Three states
of the phage were the "infective phage" or mature
phage, the "vegetative phage" (non-infective form
inside the host, recently identified by Doermann
as in the "eclipse phase" of reproduction), and the
"prophage" (distinct from vegetative phage in its
permanent state in lysogenic bacteria). He argued
that these states were essentially universal for all
phages, even though a few variant results suggested
different states between large (T2) and small (T3)
phages. The transitions between these states, according to Delbrück, had been "enormously clarified by Hershey's remarkable discovery that the in-
fecting phage behaves like a microsyringe." This now
famous result accounted for the non-infectivity of
the vegetative phage, certain antibody inactivation
kinetics, and that DNA is probably responsible for
the genetic continuity of the phage. The maturation
of vegetative phage to mature phage remained a
puzzle, with many opaque immunological experiments, but to Delbrück, "The simplest way out of
this dilemma is to assume that vegetative phage is
[pure] DNA; that upon maturation it differentiates
a protein coat; that upon premature lysis the DNA
slips out from the immature particles; and that only

the coats are recognized as phage-like material by
the techniques so far employed. It should soon be
possible to decide whether this conjecture is any-
where near the mark." . . .

When Delbrück reviewed the transition from
prophage to vegetative phage, he summarized the
various treatments that provoked this transition
("induction"). He noted that UV, x rays, nitrogen
mustards, and hydrogen peroxide all were inducers
of phage production from lysogenic bacteria. In his
usual clear and forthright language, he wrote: "Induction of lysogenesis is thus shown to belong to
the same class of phenomena as mutation and carcinogenesis, and shares with it [sic] our complete
ignorance of its true nature." . . .

He identified phages as belonging to one of
two general types: temperate and virulent. Temperate phages are those that can make a transition to a
prophage in some of the cells of a susceptible host,
whereas a virulent phage cannot. Recent experiments indicated that mutations of temperate phages
that could render them virulent were common, suggesting that the property of temperance was under
the control of the genes of the phage.

Two other topics that Delbrück identified as
important to the field of phage research were the
recognition of the similarity between phage attachment to bacteria (involving specific "receptors") and
the interaction between antigens and antibodies and
the importance of understanding the biosynthetic
pathways for phage multiplication, being studied
with the newly available radioactive tracer methods
of the biochemists.[18]

Delbrück's summary of the Royaumont meeting was circulated in the phage community—at least to the recipients of the *Phage Information Service* newsletter—and it served as one more way he established phage dogma and defined the research agenda of the APG. By 1952 the linkages between the APG and the Francophone phage researchers in Paris, and to a lesser extent those in Brussels and Geneva, had become well established. The unusual importance of a meeting such as the one at Royaumont can be appreciated by noting the lengths that European phage workers went to in order to participate: Grete Kellenberger (1919–2011) and Eduard Kellenberger from Geneva even sold their cherished piano to raise travel funds to attend. Frequent exchanges between Paris and Caltech, based on both personal relationships and similar research styles, flourished; people, information, and material circulated in these exchanges.

It seems remarkable that only seven years after the end of World War II European researchers had returned to their labs, recruited and educated bright young students, and produced important scientific results. Some have suggested that the peculiar position of the Pasteur Institute in French science contributed to that institution's leading role in this renaissance in biological research. Less bound to academic traditions, less dependent on government bureaucracy, and historically linked to "overseas" sister institutions, the Pasteur Institute was able to be flexible when needed and open to global connections when available.[19]

There were distinctions, however, to be noted between the French and American cultures, and even between the laboratories. McCarthyism raged in the United States while in France Gaullism and Algerian independence were consuming French intellectual circles. The young scientists at the Pasteur Institute, including many who had been involved in the resis tance, were probably more engaged in politics than their American colleagues. Gunther Stent, who was in Paris from 1951 to 1952, later recalled that "the political attitudes of the Pasteur Group were more or less those of the *Nouvel Observateur*, i.e., non-Communist left, and fiercely patriotic French (*L'Algéne: c'est France!*). Only Monod was openly anti-Gaullist; the others—Wollman, Jacob, Lwoff—were crypto-Gaullist, admirers of the general in their heart of hearts, I think. Everyone was strongly anti-American. Philosophy was not discussed much, except that ethics was viewed from the Existentialist viewpoint. (*'Il faut c'engager.' 'L'existence précède l'essence'*) [We must commit. Existence precedes essence]. The break between Camus and Sartre was viewed with regret; both were regarded as *brave types*."[20] The anti-Americanism, however, must have been non-specific, because strong individual friendships developed between the French and American phage workers during this time.

A third and, in a way, more individual tie between the APG and the Francophone phage researchers was the Swiss physicist-biologist Jean-Jacques Weigle. In 1948 Weigle resigned his physics professorship in Geneva and joined Caltech as a research scientist associated with Delbrück. He was a relatively young, independently wealthy electron microscope expert with an American wife, and he had just experienced a heart attack at age 47. He decided to shed the lifestyle of a European department head in favor of the more relaxed environment of a California research laboratory. His background as a successful physicist and his open temperament made him ideally compatible with Delbrück and his group of phage biologists. Weigle thrived at Caltech, where he studied the temperate phage λ while the rest of Delbrück's group focused on virulent T-phages. Among his notable discoveries was the phenomenon of restriction and modification of phages (with Giuseppe "Joe" Bertani) and error-prone DNA repair (Weigle mutagenesis) induced by radiation damage to the host bacteria prior to phage infection.[21]

Beyond his scientific contributions, however, Weigle promoted strong connections between the APG and Geneva. As later noted by Joe Bertani: "Jean Weigle came to Caltech in 1949, but he used to spend summers in Geneva in the biophysics lab that was part of his former physics department, so that he had quite a bit of influence on the direction of the work there and was for many years a direct connection between Geneva and Caltech." Weigle saw Delbrück's promotion of phage research through laboratory courses and small phage meetings as something needed in Europe, and in the late 1940s he applied for Swiss funds to conduct a European phage course in Geneva. Although this request was unsuccessful at the time, by the early 1950s some laboratory training in phage biology for a few students commenced. Finally, in 1963, the first freestanding European phage course was organized by Weigle's protégé, Eduard Kellenberger, at the International Laboratory for Genetics and Biophysics in Naples. "The local staff and teachers from Geneva and Pasadena could finally do what Jean Weigle had attempted to initiate more than ten years earlier! Similar courses took place in the following years in the same laboratory with other invited teachers from the USA and Geneva." Eventually sponsorship of this course (and several others) was taken over by the newly formed European Molecular Biology Organization (EMBO, 1963). The phage course model had become part of twentieth-century scientific culture.[22]

There were two other cases that defied the simple relations between the Francophone phage community and the Americans: phage research in the United Kingdom and in Australia showed a much more complex relationship to the larger phage community. It is likely that these two schools of phage research were, at times, marginal to the APG, because the focus of both the British and Australians—their framework of shared commitments, if you will—came out of the tradition of medical bacteriology and pathology, rather than genetics and physics. The early interest in virology and phage in Britain can be traced to the medical bacteriologists such as Almroth Wright (1861–1947), Paul Fildes (1882–1971), and Christopher Andrewes. And, since the leader of the Australian school of phage work was Macfarlane Burnet, a protégé of Andrewes as well as a medical virologist and pathologist, the Australian school shared many basic commitments with the British. Additionally, although Burnet had made pioneering contributions to the understanding of lysogeny in the 1930s, by the post–World War II period he had turned his attention to animal virology and immunology, for all practical purposes abandoning phage from then on.[23]

In the 1960s European science came into its own, especially in the new field of molecular biology, and phage research was recognized as a key component of this work. In the United Kingdom, the Medical Research Councils supported laboratories built around key individuals; some of these, such as the MRC laboratory in Cambridge, had started focusing on the molecules of cell structure and heredity in the 1950s. Phages were seen as useful models for both macromolecular structural research and fundamental genetic studies. Phage research in the United Kingdom, as in the United States, was initially fragmented both geographically and intellectually. The traditions established by Andrewes and Elford seemed to have no heirs, and Twort's early brush with phage left no visible legacy. Almost forgotten now were the research programs of John S. K. Boyd, a physician interested in phage therapy, who became engaged in understanding lysogeny, and Cyril Hinshelwood, whose ideas on mutation are now, justifiably, historical relics. Were it not for the later success of his student Sydney Brenner

(1927–2019), Hinshelwood's phage foray would never have become known. Brenner, however, became a phage enthusiast, albeit almost an autodidact rather than a true disciple of the APG. Returning to South Africa after his degree at Oxford, Brenner undertook a study of intracellular phage growth by developing a way to make the infected cell permeable to various additions to the medium, such as radioactively tagged precursor molecules. This approach used the bacterium *Bacillus megaterium* and one of its phages because *Bacilli* were easy to permeabilize (protoplast) compared to the T-phage host, *E. coli.*

Apparently through a chance connection, Brenner was introduced to Gunther Stent, who had just set up his lab at the University of California in Berkeley, and Brenner arranged a visit to Berkeley in the summer of 1954. Later, Stent gave his impression of Brenner to Stent's dissertation adviser, Frederick T. Wall (1910–2010): "Brenner suddenly appeared on the scene at CSH in the summer of 1955, when hardly anyone had ever heard of him (I was one of the few who *did* have some advance notice of his arrival, because not long before my brother met Mrs. Brenner on a London-Capetown Steamer). Since then Brenner has rapidly become a central figure in the avant-garde circle of molecular biologists, as suggested by the fact that two of the very best American workers, Benzer and [George] Streisinger, are spending their sabbatical leaves with him in Cambridge [where Brenner had relocated from South Africa] this year."[24]

Although Paul Fildes, one of Britain's most famous microbiologists of this period, was interested in phage, his efforts along these lines seemed limited to supervising a few students in their own phage projects. Francis Crick (1916–2004), in collaboration with Brenner, was famous for the use of the T4rII phage mutation system of Seymour Benzer to elucidate the triplet nature of the genetic code, although Crick was not, by most accounts, a phage biologist, but rather "a phage user"; he was one of several phage stars who did not participate in the Delbrück celebration in 1966 that resulted in the *Phage and the Origins of Molecular Biology* festschrift.[25]

Scientists taking up phage in the 1950s and 1960s in Europe, especially, developed a different relationship to the APG than those who had coalesced around Delbrück, Luria, and Hershey in the 1940s. They were able to build on the foundations of the APG yet assert independence from its culture and agenda. In his historical reminiscences in 1995, Eduard Kellenberger from Geneva provided a European perspective on phage research at this time. He noted that biochemists seemed to be more concerned with phage research in Europe, while it was the physicists in North America, for the most part, who championed phage work. This distinction may not be fundamentally disciplinary so much as temporal: phage research in America developed a bit earlier than in Europe, and the tools used for phage studies derived from physical methods of radiobiology. After World War II the newly available radioisotopes from the atomic bomb project gave new life to biochemistry with the new "tracer technology." Biochemists became energized to attack many problems of biosynthesis that had been awaiting better methods, and they gradually started to replace the more indirect physical approaches of radiobiology with its survival curves and target theory approaches. From the 1960s onward, phage research took a "biochemical turn" in both North American and Europe. During the 1960s, a steady stream of young American phage workers went to Geneva to participate in the European biochemical program. The problems related to intracellular events during the eclipse phase of vegetative growth, the mechanisms of phage multiplication, and gene expression all began to yield to biochemical approaches.[26]

Cognizant of the dangers of counterfactuals, we might

wonder, however, what might have been the course of the APG without these rich, reciprocal interactions with, especially, European phage researchers. The linkage of lysogeny with cellular growth and the role of genes in embryology, a particularly European focus, was more or less forced on the APG by the progress of the Francophone phage community. Electron microscopy, developed in interwar Germany, was first pioneered as a tool for studying phage and other submicroscopic agents in Germany, Switzerland, and France, but it was developed independently as a method useful to microbiologists in the United States and was pioneered as a tool for phage research by Thomas Anderson, an early member of the APG.[27] Conversely, the dominant role of physicists and physical chemists in the APG complemented some of the more biologically oriented European laboratories, which adopted the framework of shared commitments of the American phage workers in the APG. We also should not overlook the allure of Continental culture on Americans, especially younger, adventurous men and women who were eager to explore the new world of molecular biology and, at the same time, discover the Old World.

12

Laboratory Life
Social Commitments of the
American Phage Group

I n addition to the intellectual commitments of the APG, there were shared social commitments: beliefs and values that affected laboratory life, community norms, behaviors, and ethics. These social guidelines were rarely explicit, usually established by example, and more subject to individual interpretation than the intellectual standards of the APG. But like its intellectual commitments, APG's shared social commitments reinforced group cohesiveness and fostered community building.

Probably the most widely recognized social norm with respect to its scientific practice was the openness and expected generosity among members of the APG. The quite transparent and trusting communication of preliminary experimental results was a hallmark of the APG. The obvious penalty for unfair use of another's work would be "excommunication" and ostracism. It was made clear to the initiates that there was a fundamental principle related to sharing of information as well as experimental materials, such as strains and mutant phages:

the reciprocal obligations *to give when asked, but not to ask so as to compete with the person being asked.* Another characteristic of the early APG was a certain egalitarianism among all but the three founders, Delbrück, Luria, and Hershey (who were accorded near mythic respect). Young graduate students and senior faculty members alike were subjected to, and withstood, the same withering criticisms at phage meetings and seminars. Delbrück's famous "I don't believe a word of it" was delivered equally to nearly everyone at one time or another.[1] Hershey was known to ask apparently "dumb" questions to give the speaker a second chance to clarify a point that had baffled the audience, most of whom were too timid to challenge. More than once a poorly prepared speaker was stopped in midsentence and told to come back tomorrow after the talk had been properly prepared. This was indeed done; the speaker learned a valuable lesson and was, otherwise, none the worse for the experience. These scientific critiques were part of the culture, a way to help everyone do better work. As Jim Watson put it, this is the way "good and bad science got sorted out."[2]

Science, and certainly the APG, has its own set of subcultures. Delbrück observed: "I found out at an early age that science is a haven for the timid, the freaks, the misfits. That is more true perhaps for the past than now. . . . Every one of the persons there [Göttingen, where he was a graduate student] was obviously some kind of severe case."[3]

One of the persistent myths about science is that it is "cold and impersonal," dealing only with test tubes and chemicals. Like many myths, it is not true: science is intensely social—trading ideas, arguing about the beliefs of the day, sharing and collaborating in ways large and small. As the scientist-philosopher John Ziman notes, science is "public knowledge"; it requires an audience. Delbrück understood, consciously or unconsciously, that the APG, in order to thrive, needed a social culture as well as an intellectual culture. He once claimed that he scheduled the phage meeting to conflict with the annual meeting of the Genetics Society of America so that people had to choose their loyalty (phage genetics or fruit fly genetics as he saw it). Delbrück was famous for his group outings, social gatherings with play readings and dramatizations, camping trips, and dinner parties, a forerunner of today's corporate team-building mania.[4]

While scientific culture is certainly distinct in certain ways, it is also part of the mainstream culture, absorbing and rejecting parts, consciously and unconsciously. The APG was forming during a tumultuous time in America: postwar changes in attitudes about women in the workforce; anti-communist "McCarthyism" coupled with the "lavender scare," which labeled gays and lesbians as security risks;[5] then later, the incipient sexual liberalization brought on by the work of Alfred Kinsey (1894–1956) and more available contraception; new federal patronage of science; shock at the Sputnik success of the Soviet space program; and expanding enrollments in colleges and universities, to mention some of the major social forces. These currents in American society all had their impact on the social structure of the APG.

In science and in the larger culture, the "proper" roles of male and female were discussed openly as well as contested indirectly. The issue of gender in science had its representation in the APG, too, and warrants serious consideration. The shared intellectual commitments central to the development of the APG are, at least on their surface, culture-neutral—that is, they don't seem to carry specific cultural or social assumptions. They purport to be beliefs that are not peculiarly North American, Asian, or European, neither male nor female, gay or straight, black or white. Although one might dig deeper and find, for example, that the belief in the invariance of physical laws with regard to time and place might be considered "Western" in a traditional sense, for the most part, the shared commitments of the phage group can be thought of as "scientific," relatively free of obvious cultural or gender influences. Still, we can inquire about the social and cultural values and beliefs of the APG and the ways they influenced the subsequent directions of the discipline of molecular biology. As with science itself, deeper understanding is dependent on asking the right questions. This chapter takes on the challenge of illuminating some of the social and cultural commitments, the "framework of shared social commitments," of the APG. General community norms, attitudes, and actions are probably the most important and relevant aspects of such an analysis. Individual acts and attitudes can be important as indicators, and sometimes as key determinants of community behavior, but often they signify only individual diversity, good or bad. To the extent possible, I will avoid a simple catalog of the latter and will strive to contextualize them instead.

The first caveat is to distinguish between group and individual commitments: individuals diverge from each other in specific ways and such idiosyncrasies may or may not be relevant to the *shared* commitments of the group. Viewpoints (biases?) may be shared, tolerated, rejected, or ignored. Second, every small community such as the APG is also part of the larger contemporary culture in which it functions, and it is not immune from such influences, great or small. Data for a small, nascent discipline such as the APG are hard to come by; social beliefs and practices are not often discussed and recorded as explicitly as are scientific beliefs and practices. Indirect evidence, personal stories, and inferences are often the only way to understand such cultures. More anthropology than history.

A major component of the social commitments of any group is related to gender and sexuality, and scholars have used these keys, along with race and class, to probe the cultures of science. Is physics a discipline that is organized by patriarchic principles? What would a "feminist mathematics" look like? How isolated is "scientific culture" from so-called mainstream culture? It seems helpful to distinguish between gendered *content* and gendered *practice*. Are the "content questions" posed, the aspects of nature deemed worthy of attention, the methods of their study, and so forth influenced by gender? Since science has evolved in a predominately male culture, we rarely have good counterfactual cases to approach such questions directly. We have only inferences and speculation. In some cases, such as the biology and medicine of sex differences in the nineteenth century, it would seem that, indeed, the *content* of the science was rather strongly influenced by the gendered assumptions of its practitioners. In the case of phage biology, gender (and class and race, to the extent we have any evidence) influence seems more obscure. In contrast to the social influences on the *content* of science, the social commitments of the *practice* of science are clearer. Assumptions, based on gender, race, and class, about social and professional roles, about suitability for certain types of experimentation, about innate talent (or lack of it), about theory versus experimental work, and the like have all been documented and exhibited by nearly every scientific discipline.

How can one assess the development and importance of the APG and its role in the disciplinary origins of molecular biology, yet still examine the ways both individuals and community cultures took account (or did not) of sexuality and gender as important components of scientific life? It would be too easy to lump the APG and the nascent discipline formation of molecular biology into the general story of "women in science" that has been well characterized for at least four decades by pioneering scholars of science and gender such as

Margaret Rossiter, Pnina Abir-Am and Dorinda Outram, Londa Schiebinger, Sandra Harding, Bert Hansen, and others.[6] The community of phage researchers in the years between 1940 and 1960 was enmeshed in the social context of science of its time. To discuss that context in generalities would be only a confirmatory story and would add little to what we already know, but the nascent APG was developing its own culture, both in terms of the practice of science and in terms of its social structure. The historical record provides traces and shadows of this culture that give a glimpse into that early community.

Although there are accounts and remembrances of women phage scientists, the paucity or actual absence of any specific references to diversity based on ethnicity, class, or sexual identity is evidence in itself. The heterogeneity of class and national background, so long as it encompassed only Western cultures, was taken as a given in the immediate postwar period with the influx of refugee scientists and returning veterans. America was more mobile and more open to "foreign" ideas than earlier. However, few Asians, Asian Americans, African Americans, or Indigenous Peoples were able to take up academic research such as phage biology or genetics, given that they had not yet started to emerge from the high school and college training grounds needed to participate. During the formative years of the APG, academic positions were still mostly populated by secure and financially well-off individuals, who were able to afford advanced education as well as tolerate low to non-existent salaries.[7] Even more invisible were LGBTQ+ scientists, who faced both social and legal discrimination if they publicly acknowledged their identities.

There is an interesting historical happenstance, unique to the APG, that needs further illumination with respect to sexual diversity and identities. In the late 1950s, at Indiana University in Bloomington, both the APG and the Kinsey Institute for Sex Research were thriving, both outgrowths of the same community of biological scientists. Astoundingly, at least in the phage literature, there seems to be a nearly total lack of interest in or recognition of the work of Alfred Kinsey, work that was going on literally down the hall. In his autobiography, Salvador Luria, a member of the department of microbiology at Indiana University at the time, briefly and dismissively mentions the fame of the Kinsey project: "Contributing to Indiana's veneer of sophistication was the presence of Alfred Kinsey, the entomologist turned sexologist and a rather prepossessing personality on campus. Anecdotes, mainly apocryphal, about his Institute of Sex Research abounded. But the very fact of Kinsey's presence and that of his co-workers influenced campus attitudes toward sexuality. . . . If our society lost its illusions about chastity after Kinsey, improbably Bloomington, Indiana, was the first to lose them."[8]

Women in the APG

First, let us consider the simple representation of women in the community of early phage workers. Beyond the numbers, what roles did these women play, what were their contributions and influences, and what internal and external factors impinged on their activities and lives as scientists?

As described earlier, there were several crucial strategies involved in the formation and evolution of the APG—foremost, perhaps, was the phage course, a key gatekeeper for the recruitment, indoctrination, and acceptance into the field, as envisioned by the nucleus of founders. Consequently, the phage course is a reasonable place to start our inquiry. The first year of the phage course, 1945, was unusual in that the course was organized during the final year of World War II. Young men, but rarely young women, were still in active military service, so the age and gender distribution in that first year may have been affected by this constraint. There were only six participants, two of them women, so the "law of small numbers" makes strong generalizations impossible. However, aggregating the phage course alumni of the first ten years (1945–1954) shows that there were 38 women out of a total of 145 students (26 percent). In the same period, there were 16 international visitors, not counting international scientists based in the United States (11 percent). It is not possible, using available data, to identify individuals who might have self-identified as other gender-diverse categories. At least 7 of the 38 (18 percent) female students went on to have (identifiable) academic careers at major research universities: Martha Barnes Baylor, Harriet Taylor Ephrussi, Ruth F. Hill (1917–1973), Margaret (Peggy) Lieb, Marguerite Vogt, Helen Van Vunakis (1924–2018), and Birgit Vennesland (1913–2001). Collectively, however, the women selected for the phage course in these early years tended to be younger and less well-established than their male fellow students, who were, by and large, mid-career or senior scientists. Did this reflect a bias in favor of early career women to bring them into the field and promote their scientific status, or, alternately, did this occur because they were being trained in skills to be transferred back to their home institutions as lab assistants and research technicians? A significant fraction of the male scientists in the phage course seemed to be senior scientists from the physical sciences, who did not do any significant phage research subsequently. It might appear that they were simply indulging their curiosity about this new enthusiasm of some renegade, but reputable, colleagues in the community of atomic physicists (for example, Feynman).[9]

Whatever the individual motivations of each participant (and we have scant complementary information about applicants who were rejected), it is not possible to know for certain whether the significant fraction of the phage course initiates who were women represents a proactive encouragement of women scientists in the nascent culture of the APG, or, conversely, a random product of the available pool of applicants or a result of institutional sponsors' desire to use the course (in contrast to the stated aims of Delbrück and Luria) as a technical training ground for lab assistants. But in any event, a significant number of women were given the opportunity to explore phage in the approved way.

Founder Effects and Diversity

Since discipline formation usually involves a very small nucleus of individuals, a pattern that holds true for the APG, it is interesting to consider what population biologists call "founder effects"—that is, the role played by the few founders who have outsize influence on the directions, both intellectual and social, of the community of their followers. This analysis provides an individualized account of the origins and development of specific social practices, including gender roles and attitudes, that became characteristic of that community.

Such founder effects have been described for the field of x-ray crystallography that sprang up around William Bragg and Lawrence Bragg (1890–1971), which had a significantly higher proportion of women leaders than most other physical sciences in the twentieth century. This has been attributed to the support given to women crystallographers by the Braggs as well as their protégé, J. D. Bernal. The mentorship of a few founders set the course for the subsequent development of the discipline as a community that provided opportunities for women scientists that were rare elsewhere. As Julia Sanz-Aparicio observed: "The early days of crystallography are, thus, replete with excellent female scientists. Since X-ray crystallography was born during the second decade of the 20th century, and only developed after World War I, the field is relatively new and perhaps more free of traditional prejudices. The early appearance and welcome of a few brilliant women into the laboratories of the Braggs was crucial. Many of the co-workers of these women, especially J. D. Bernal, later carried on the tradition of inviting women students and colleagues into their own laboratories."[10]

In the case of the APG, we have the opportunity to examine such founder effects and to see how gender factored into the culture of this new and evolving discipline. Data that might be informative include the roles played by women (and sexual and gender minorities, to the extent that we can identify them) in the small but growing community of phage workers. We know who took the phage course and who attended the phage meetings. Beyond the simple numerical information, however, we can ask, what was the mentorship of women like and what were the gendered experiences of both women and men in the APG? In a few cases we have hints from personal accounts and recollections.

Delbrück, probably the most forceful personality of the three main founders of the APG, was unusual in many ways but unique in that his early professional work involved close collaboration with two of the most renowned female physicists in the German-speaking world: the eminent nuclear scientist and Nobel Prize recipient Maria Goeppert Mayer (1906–1972), a fellow graduate student under Max Born in Göttingen, and Lisa Meitner, the co-discoverer of nuclear fission, for whom Delbrück would later work as her assistant in her Berlin institute. One could not have had a more impressive firsthand exposure to talented and successful women scientists. By most accounts, Delbrück was supportive and generous toward his female associates, but contemporary documents, in keeping with the attitudes of this era, rarely ventured explicitly into matters of sex and gender. We are left to infer such understanding from fragmentary gleanings, offhand remarks, and occasional recorded events.

In an essay intended to be amusing, Delbrück's wife, Manny, early in their marriage, described Max as a new species, Homo Scientificus: "easy and interesting to observe, but difficult and perplexing to understand . . . He does not . . . delight in flaunting elegantly before the female to catch her eye [rather] he brings her a little gift such as a bundle of bristles or a bright piece of cellophane, which she accepts tenderly, and the trick is done." Homo Scientificus is content "so long as he can spend most of the day sitting in the sun and rummaging among his strange possessions." Martha Baylor and Phyllis Margaretten, the two females in the original phage course organized by Delbrück at the Cold Spring Harbor Laboratory two years earlier, were both chemists at the University of Illinois, Baylor as a post-doctoral researcher and Margaretten as a graduate student. Baylor later had a distinguished career in biology as a faculty member at the University of Michigan and later at the Woods Hole Marine Biological Laboratory and then at SUNY Stonybrook. Margaretten received her M.S. in chemistry at Illinois; her son later reported: "As was perhaps typical of the period, her scientific career ended with marriage and motherhood though she continued to follow molecular biology and related scientific developments closely. Later she studied and did research in nutrition, receiving a Ph.D. from the University of Maryland and then working for the advocacy group, Center for Science in the Public Interest." Gunther Stent, Delbrück's ardent admirer and protégé, described his first meeting with Delbrück when Stent was still a graduate student in chemistry at the University of Illinois: "Delbrück, by the way, seems to know both Martha and Phyllis quite well. He referred to Lester Machta [husband of Phyllis Margaretten] as

Mr. Margaretten.—I think I am going to like that Biologist crowd; they don't seem to be deadheads like the chemists."[11]

That Delbrück was a complicated person, even beyond the understanding of his wife, can be seen in an episode recounted by one of his early post-doctoral students and later a professor at the University of Southern California, Margaret (Peggy) Lieb. She described Max as "much respected by those in the church [the APG], and despised by some of those outside, because he could be ruthless in dispensing with the ideas of inferior scientists."[12] After working with Delbrück at Caltech, Lieb joined the Pasteur Institute in 1953–1954, working with Jacob, Monod, and Lwoff on phage. She shared a small lab with another young visiting American phage worker, Aaron Novick. Novick was known to be insecure, self-described as a melancholy Jew, and as a defense he assumed an air of superiority. He was particularly high-minded about his resistance to the loyalty oath requirements brought on in the United States by McCarthyism. When Delbrück paid a visit to the phage group at the Pasteur, he arranged an April Fool prank on Novick: he had an official-looking document delivered to Novick's apartment indicating that he would be required to sign the loyalty oath to keep his U.S. salary and position at the Pasteur Institute. Novick was genuinely anguished, fearing for his livelihood and his wife and young child if he did not sign, yet morally and publicly committed not to sign. By the next night, April 1st, he had already made plans to leave Paris and seek another job when the prank was revealed at a dinner party hosted by Delbrück, and Delbrück, very cruelly in Lieb's view, embarrassed Novick to the amusement of his French colleagues. His conduct clearly established Novick's status as an outsider. Lieb, feeling like an outsider herself, described her sympathy for Novick, even though he had treated her with aloof superiority upon his arrival in Paris as her nominally equal colleague and office mate.

In contrast to Delbrück, the second of the three founders of the APG, Salvador Luria, was more politically active,
more open about his personal beliefs, and more publicly in-
trospective about his ethical and moral developments. In his
autobiography he describes his marriage to Zella Hurwitz, a
professor of psychology, as a "model couple in modern house-
keeping. We alternated monthly in shopping and cooking and
in assuming responsibility for other household chores. This
exercise . . . was in part an assertion of our shared feminist
convictions." Luria had a sheltered childhood, plagued by sporadic illnesses, and he developed an aversion to traditional
sports and physical activities. He "experienced fear of the bully
as well as discomfort in the company of more adult, more mas-
culine schoolmates"; even as an adult "this fear has made me
sensitive even to the smell of gymnasiums, the coarse effluvi-
um of masculinity that probably rouses the energies of ath-
letes while it nauseates me. It has made me uncomfortable in
the presence of people whose personality suggests the possibility of physical violence. Perhaps for that reason I have always been more comfortable in the company of women than
of men." Similar to Delbrück, Luria was no stranger to successful female scientists during his formative years. During his sojourn in Paris, immediately before his emigration to the United
States in 1940, he worked in the Radium Institute, famously
founded by Marie Curie, where one of Luria's colleagues was
Curie's daughter, Irene Joliot-Curie (1897–1956), a laboratory
director and Nobel Prize recipient in her own right.[13]

Laboratory Life

Although it seems clear that women participated, often as respected and seemingly equal colleagues with men in the APG,
we can still ask about the nature of that "equality." As Hilary

and Steven Rose described Rosalind Franklin's (1920–1958) acceptance in Bernal's crystallography department at Birkbeck College after her DNA structure fame, she was welcomed by Bernal for her scientific competence, but at a cost: "In this laboratory, to be a woman scientist was to enter in an honorary capacity the brotherhood of men and thus to be (almost) above heterosexual invitation. As many women who have worked in almost entirely male enclaves know, becoming this kind of honorary man is one strategy to desexualize the environment and make work relations manageable."[14]

This situation was captured succinctly in an essay by Shirley M. Tilghman (b. 1946), a molecular biologist and university president, when she quoted a friend who described science "as a black hole, prepared to suck up whatever proportion of your life that you allow it." As Tilghman explained her experiences:

> This complete devotion to science was fostered in the culture of the 50's in which women stayed home and raised families while their husbands conquered the secrets of the universe. When women began to enter science careers in the 1940's and 1950's, they were expected to renounce any intention of having a family. This is the ultimate un-level playing field, one that persists to this day. Women have paid a terrible price for the success they have realized in the last 20 years. Study after study of all fields, not just science, document that women have forgone marriage and children for their success.[15]

In a community of young scientists with the revolutionary fervor of the APG, science seemed to be organized as a full-time life commitment, indeed a black hole for effort. As the sociologist Barney Glaser observed in 1964, scientists face "a life of comparative failure" when they model themselves on the heroes of their fields.[16] In science, one answer always seems to lead to new questions, a never-ending series of failures to achieve completion, a black hole that can be both exciting and debilitating.

The APG, to the extent we can know it, promoted certain communitarian beliefs, limiting competitive individualism. This cooperative communitarianism is seen by some as a "female trait" that contributes to a supportive laboratory environment. A second-generation phage worker, Ann Skalka (b. circa 1938), recounted her happy and positive experience at Cold Spring Harbor Laboratory as a post-doctoral fellow in Hershey's lab. She described her colleagues as "a diverse, creative, and wonderfully congenial group of young phage researchers" (Ruth Ehring [1930–2018], Gisela Mosig [1930–2003], Eddie Goldberg [1935–2015], Mervyn Smith, and Betty Burgi).[17]

Cold Spring Harbor Laboratory, on the other hand, retained characteristics from a distinctly male tradition: the living quarters modeled after boarding school and military life— spartan rooms, little privacy, and communal shower and toilet facilities. Delbrück, likewise, promoted a social structure for his group at Caltech in terms that might be seen as reinforcing typically male comradery, perhaps viewing the women in the group as "honorary men." His frequent command performances with his lab group (and even distinguished foreign visitors such as André Lwoff) on camping trips to Joshua Tree National Park in the desert east of Los Angeles are legendary. Luria, however, would not visit Delbrück in California unless he was guaranteed immunity from these desert camping ordeals. Laboratory social life and laboratory scientific life were inevitably intertwined; one avoided social events at peril to one's scientific status.[18]

We can reframe the question in another way: how do one's social life and scientific life interact? Bert Hansen has suggested that the laboratory life of a scientist can be strongly affected by the social support (or lack of it) that one has outside the laboratory. A nurturing home life with a sympathetic partner could be conducive to a productive and creative research career, while a lonely and stressful personal life might spill over into one's intellectual life in the laboratory.[19] In the era during which the APG was forming, same-sex and transgender relationships carried the additional burden of secrecy in the face of both legal and social discrimination. Since there is no easy demarcation between the scientific and the social, it is not surprising that social relationships often grew out of scientific relationships, and the APG was no exception. Charles Yanofsky (1925–2018) at Stanford was said to have "a strict rule that there should be no 'affairs' in the lab, and in fact had asked a post-doc to leave because he (or she) was playing around with one of the post-docs of the opposite sex."[20] Yanofsky's caveat notwithstanding, the APG seemed to have its share of collaborating couples. Mention has already been made of the work by Esther and Joshua Lederberg on lambda phage. Ethyl and Irwin Tessman, both faculty members at Purdue University, met as students at Yale, and they became authorities on the small phage, S13. Elizabeth (Betty) Bertani and her husband, Giuseppe, met in the lab at the University of Illinois; after Betty obtained her Ph.D.—as the first woman to do so in the biology department at Caltech—they both went on to collaborate on problems of lysogeny as faculty members at the University of Stockholm. Other such "phage couples" include Cynthia Lark (1928–2005) and Karl (Gordon) Lark, Grete and Eduard Kellenberger, and Helga Harm and Walter Harm (b. 1925). These academic couples were consistently considered "pairs" in an equal sense and they published both together and independently. They were so frequently referred to simply as "the Tess-mans" or "the Lederbergs" that one might think of this group of scientists as "the plurals."

Although such positive outcomes suggested progress in gender equality, other, less happy examples exist as well. Martha Chase, who famously collaborated with Hershey on an experiment that still carries her name eponymously, and who worked later at Oak Ridge National Laboratory and at the University of Rochester with Gus Doermann, enrolled for a Ph.D. at the University of Southern California to study under Joe Bertani. Chase had a short-lived marriage to Doermann's student Richard Epstein (1929–2011), briefly going by the surname Chase-Epstein, and she was strongly affected by their divorce, according to her lab mates at USC. Although she eventually received her doctorate under the direction of Peggy Lieb (after Bertani left USC for Stockholm), Chase did not continue in science, and she struggled with personal issues for the rest of her life.[21]

As if to provide textbook examples of Delbrück's diagnosis that scientists all were "some kind of severe case," two prominent APG members, Jim Watson and Gunther Stent, wrote rather strange, exhibitionistic memoirs. Watson's late-in-life autobiography is titled *Genes, Girls, and Gamow: After the Double Helix,* peculiarly alliterating his key focuses. A similar fixation was on display in a lecture he gave at Yale in the 1990s dwelling mostly on his rocky love life as a young academic. When I observed the disappointment of several students that Watson did not give the scientific lecture they expected and instead talked rather insensitively about the women he dated, a senior colleague who knew him very well quipped, "He just described his hunt for a Jim-Watson-resistant mutant." Stent's autobiography, *Nazis, Women, and Molecular Biology,* another rather bizarre title, overflows with texts of his youth-

ful love letters and graphic descriptions of his various roman-
tic exploits. Maybe Delbrück is correct in his observation that
science provides a refuge, a place in society, for individuals who
are "cases," some with arrested development of certain social
and interpersonal skills, never quite resolved even into later
adult life.[22]

In addition to the APG's tolerance for individual diversity of behaviors, compared with the community of biological
science as a whole, the fledgling APG developed a social culture and atmosphere distinctly more relaxed, more informal,
less hierarchical, and just plain more fun than the older, staid
communities in biology. Jackets and ties were almost forbid-
den at phage meetings, and shorts and sandals were frequently
the dress code for speakers there. It was not unusual for an eve-
ning speaker to appear at the podium with a can of beer from
the basement bar still in his or her hand. In the classrooms and
laboratories of colleges and universities, students differentiated the professor of molecular biology from the professor of
comparative anatomy by demeanor and dress. Undergraduates
addressed professors by their first names without any thought
of disrespect. Open challenge of a scientific assertion was not
seen as a challenge to authority but as part of the scientist's task
to sort out warranted belief from weakly held ideas. This informality, sometimes rather forced, came to characterize the social
structure of the APG and its offspring, molecular biology.[23]

I experienced this assumption of informality in action
when I first joined the laboratory of Cyrus Levinthal at MIT in
the fall of 1967. Coming from the midwest and being unfamiliar with the culture of MIT (or Bostonian New England) at the
time, I arrived for my first day in the lab wearing a jacket and
tie.[24] Seeing that everyone was wearing the de rigueur jeans and
"engineer boots," I showed up the next day as informally as I
could muster, along the same lines as my lab mates. Later in

the year, however, during a lab celebration related to the potential ending of the Vietnam War, canvassers for a keg of beer for the cold room avoided asking me for a contribution. Years later, when reminiscing about those events, I asked a former lab mate, by then a good friend, why I was not asked to contribute. He said that on the first day I arrived in the lab in jacket and tie, with an M.D. degree from the midwest, I had been immediately typecast as a political conservative, and all my "coastal liberal" colleagues assumed I would not be inclined to join their revelry (all untrue). As subsequent events showed, many of the most politically active members of the biological community came from those who had been associated with the APG and identified as molecular biologists.

The Invisible Hand

A significant influence on the social structures of science is the nature and source of patronage. Once that might have been aristocratic favor or royal grants. By the early days of the APG, science relied on private philanthropy and, later, government largesse. Like the guiding hand of Adam Smith, patronage in science is often invisible or at least murky. Yet material, and sometimes intangible, support is crucial for science and the way it is practiced. Who gets funding and for what kinds of work are all important, especially in the early days of new directions. "As the twig is inclined, so grows the tree." As others have shown, a major source of support for biological science in the interwar period was the Rockefeller Foundation, an institution with a strong commitment to vertical integration and top-down direction.[25] The leaders of the Rockefeller Foundation took a hands-on approach to their philanthropy and were intimately involved with the people and programs they sponsored. This allowed for flexibility and timeliness of support, but

it also resulted in insular decisions based on close personal relationships, suppression of heterodoxies, and reinforcement of
founder effects. The program officials for the Rockefeller Foundation, the National Foundation for Infantile Paralysis, and the
American Cancer Society often had direct and close contacts
with favored scientists whom they sustained both formally
and informally. This was the essence of the "old boy network"
that stabilized and perpetuated both good and bad aspects of
science patronage. One official might be open to diversity and
innovation and another only interested in backing "tried and
true" endeavors. Again, with small organizations, founder effects could be pronounced.

Unfortunately, we have scant actual data about the decisions and attitudes involved in this kind of patronage. Statistical data, too, are unreliable because of the small numbers of
individuals operating in the APG in its early period. Still, the
general impression from the sum total of data is that patronage flowed to a few of the principal phage researchers, who
then distributed it to their junior colleagues as they saw fit,
with the tacit and sometimes explicit approval of their philanthropic patrons.

With the postwar expansion of federal science policy and
financial support, patronage of science changed dramatically,
so that by the 1960s, private philanthropy was being dwarfed
by federal funding. Federal support for the APG came from
several sources: the Atomic Energy Commission, reflecting the
radiobiological origins of phage research, the National Science
Foundation with its broad mission to bolster the sciences,
and, a bit later, the National Institutes of Health, as the phages
became identified as model viruses. Importantly, the nation
recognized the need to educate more scientists in the post-
Sputnik panic of 1957 and started programs to strengthen training for young scientists as well as to expand and improve uni-

versity science departments. The government's science support was directed by several competing forces: political patronage for regional distribution of funds, outside boards of expert advisers, the so-called study sections, and a larger, more diverse community of government officials who charted priorities and goals. Open advertisement and announcement of available grants, competitive fellowship programs, and more transparent evaluation of proposals became the norm. Little by little, patronage became a bit more diverse, more open, and more equitable, and female scientists began to take their place in the power structure of science.[26] Recognition of the need for affirmative action programs to help achieve full participation of all talented individuals was codified in U.S. law starting in 1961, but even by the early 1970s concrete progress was still slow.[27]

In retrospect, it appears that the 1960s represent a tipping point when two factors joined to change the landscape for women in science. One was the change in patronage to federal support, which was legally more egalitarian, and the second was the change in reproductive control available to women. Oral, hormonal contraception and the Supreme Court decision in *Griswold v. Connecticut* legalizing other forms of contraception placed reliable fertility planning clearly in the hands of women for the first time. But still, that black hole of science beckoned.

13

Maturation and Assimilation
From Phage to Molecular Biology

The early shared research commitments of the APG revolved around two ideas: that "theorizing" was primary and that target theory and other "black box" input-output approaches would provide understanding of invisible objects and processes. Simple phage experiments—such as one-step growth curves, rates of inactivation of plaque-forming activity by x-rays, and burst-size measurements from various phage combinations—all fit nicely into this program. Indeed, much about phage became clear in the decade of the 1940s through such experiments. About 1950, however, troubling results kept cropping up. It turned out, for example, that the survival fraction of a sample of phage treated with x-rays depended on the ratio of phage to bacteria (the multiplicity of infection) used to assay the survival. Simple target theory assumed that "dead is dead" but this phenomenon called multiplicity reactivation suggested that dead phages could somehow cooperate to reconstitute living phages. Clearly something was not right. The phenomenon, first observed by Luria, promised to offer new insights into what was going on with phage growth during the intracellular eclipse period. Also puzzling was the fact that phages killed by ultraviolet light could be resurrected by exposure to visible light, an effect termed photoreactivation. Bacteria killed by radiation did not appear to be really dead, because if they were held for a while in distilled water, or other non-growth media, they came back to life, the so-called liquid holding recovery. In target theory terms, things were not behaving properly. The target was formally changing size or sensitivity depending on post-hit conditions. The target did not "stay hit."[1]

Two other phenomena had become central to the APG research agenda as well. From early studies, initiated by Ellis even before the Delbrück era, on the nutritional requirements for phage reproduction, it was found that some phages required certain "co-factors" in the medium for successful reproduction. These studies eventually settled on the observation that for phage T4 infection, the amino acid tryptophan was needed. In Delbrück's view, the "tryptophan effect" became a central problem needing a solution. For reasons no longer clear to us, he seemed obsessed with this "co-factor story," believing that it was going to be central to the understanding of phage reproduction. Numerous students were assigned to this work, which seemed to excite Delbrück a lot more than most of his co-workers. To quote Tom Anderson, who discovered the tryptophan effect, "He [Delbrück] and his students spent many enthusiastic years trying to solve the mechanism of cofactor activation. Unfortunately, the cofactor mutants have so far not been very useful in phage genetics." As it turned out, tryptophan serves to modify the phage tail fibers to promote attachment to bacterial surface structures needed for T4 entry into *E. coli*.[2]

Another research program that generated much enthusiasm was a variant of the standard target theory approach to inactivate phages, first reported by Hershey, Kamen, Kennedy, and Gest, who found that radioactive atoms of phosphorus, incorporated into the DNA of phage, would inactivate phage as the radioactive phosphorus decayed over time. This type of experiment, known as the "^{32}P suicide experiment," became the province of Gunther Stent and his students for nearly a decade; it spawned an entire generation of derivative approaches using radioisotope decay as the source of radiation in target theory experiments. As elegant and theoretically sophisticated as some of these experiments were, they did not, in the end, provide answers to the central problems of phage reproduction—namely, what are the molecular structures and reactions that are involved in the production of progeny phage from the input parental phage? That is, how does like beget like?[3]

To make matters worse, the same material showed different target theory behavior whether irradiated in the wet or dry state or depending on addition of various solutes to the medium in which the irradiation was done. Indeed, the Ph.D. thesis of one James D. Watson was on the effects of adding the amino acid histidine to the buffer in which target theory analysis of x-ray killing of phage T7 was carried out. To account for these complications, the target theory was modified to include both the "direct effect" of damage to the molecular target and "indirect effect," which was damage to the molecular target relayed by radiation products produced in the aqueous solution within diffusing distance of the actual target. This revision of the target theory undercut its principal advantage, that of ignoring the unknown chemical milieu of the living cell.[4]

As radiation biology became more intensely and quantitatively studied, bacterial strains were isolated that had altered sensitivity to killing by x-rays and ultraviolet light. Evelyn Witkin isolated a mutant of *E. coli* B that was more resistant to radiation than the normal *E. coli* B strain and called it strain B/r. A decade later Ruth Hill isolated mutants of *E. coli* B that were more sensitive to radiation, designated *E. coli* Bs-1. Even more puzzling for bacterial target theorists was the finding that the same samples of irradiated phages gave different survival curves when assayed on the sensitive, normal, and resistant strains. This phenomenon was termed host-cell reactivation (of the phage). By now, it was becoming clear that there were metabolic processes in the cell that affected the damaged phages, processes that could be altered by cell mutations. As Frank Stahl noted later, "Like most early experiments in 'radiobiology' these analyses [based on the idea of targets and hits] were fun, but not much more." Target theory was out and cell physiology was in.[5]

A key step in understanding these results was the discovery in 1964 of metabolic pathways in *E. coli* that specifically remove and repair damaged DNA. The physicists who reveled in target theory and its simple black-box logic were now confronted directly with the challenges of biochemistry. Despite Delbrück's suspicion of biochemistry, the APG would never be the same.[6]

The APG was not without new tools and ideas, however. They now had DNA. In the decade of the 1940s, the chemical nature of the gene took a back seat to the formal physics of the gene. For example, Luria's elaborate explanation of multiplicity reactivation was diagrammed with simple boxes in various combinatorial arrangements. The chemistry of DNA, such as it was, remained in the domain of the few biochemists who thought it might be as important as proteins. All this changed with the elucidation of the double helical macromolecular structure of DNA by Watson and Crick in 1953.[7] Now biologists could "see" *function* in the *form* of chemistry. The biblical mystery of how like begets like was solved by understanding the stereochemistry of base pairs. Not only chemistry but also the nascent field of information theory was invoked. Complementary base pairing provided the exactness of reproduction, but the genetic specificity to carry out the "code" in DNA required understanding "how DNA directs the cell." These new frontiers required expansion and readjustment of the framework of shared commitments of the APG and their increasingly diverse allies. How are proteins made in the cell? How are the genes and proteins related? How are such reactions in the cell coordinated and regulated? In phage terms, how is the phage growth cycle programmed? It was easy to think of the phage development cycle as analogous to the embryology and development of higher organisms: parts and assembly reactions had to be organized in time, place, and quantity. Macfarlane Burnet's idea of *Anlagen*, a plan for phage growth and reproduction, inscribed in the sequence of nucleotides in DNA, had come of age. A *genetic program* in the same sense as a computer program.

This might have been the end of the APG, but its relevance was extended by one of the contentious problems from its beginning: lysogeny. The study of temperate phages, especially phage λ, and lysogeny provided an important bridge between phage biology and the growing importance of physiological genetics, focused as it was on the regulation of gene expression.

An overarching view of genetics since its modern incarnation suggests two main themes or foci: how does like beget like? and how do genes control the outcome of life? These themes can be stated more formally: heredity and embryology. Scientists studying the mechanisms of heredity and breeding called themselves geneticists and studied how a trait was passed on from one generation to the next. This was the Mendelian program. Since the early part of the twentieth century, this was the focus of the American school of genetics of which T. H. Morgan's work on fruit fly heredity was the prime example. More prominent in Europe, since the time of Ernst Haeckel

(1834–1919), was the embryological program aimed at under-standing what controlled the development of an organism from zygote to adult.[8] In modern terminology, these research paths might be characterized as "gene replication" versus "gene reg-ulation." Even in the 1950s, these two schools of genetics were hardly on speaking terms. While it may have been the un-spoken hope of both sides that eventually their different views would be reconciled by some unifying understanding, there were only sporadic movements in this direction—that is, until the clarification of the chemical nature of the gene ushered in by structural studies on DNA, and the focus on the nature of mutations and the genetic code. The centrality of the DNA structure to clarifying gene replication and the direct connec-tion of the cell instructions coded in the base sequences in the DNA provided this unifying hypothesis. There became "trans-mission genetics" and "physiological genetics."

Two powerful genetic tools opened up experimental space to investigate this hypothesis: genetic control of metabolism and genetic control of phage growth. Lysogeny and temperate phage growth suited this program well. Because mutations in bacteria and in phage were found that changed the course of bacterial growth and metabolism, or modified the course of lysogeny, the gene as a unit of heredity and the gene as a con-troller of function became linked in one experimental system. The phage workers in Paris, in particular, led this grand unifi-cation. They employed the growth requirements of bacteria as an indicator of function, a sure marker of cellular biochem-istry. The early history of bacterial nutrition at the Pasteur Institute can be traced at least as far back as 1906 with the recognition of a strain of E. coli called mutabile by Max Neis-ser (1869–1938) and Rudolf Massini (1880–1954) in which the control of lactose utilization could be changed by genetic mu-tation. The biochemical pathway for lactose metabolism was simple and relatively easy to study in bacterial cultures, and it became a model bacterial function that was studied by Lwoff, Monod, and their colleagues in the Laboratory of Microbial Physiology at the Pasteur Institute.

In 1941 Monod discovered the phenomenon called *diauxie*, the observation that bacteria take some time to change their metabolism from using one sugar, say glucose, to another sugar such as lactose, and the Pastorians realized that this metabolic switch involves the de novo synthesis of the enzymes of the new metabolic pathway. Joshua Lederberg found that a particular enzyme, beta-galactosidase, was present in *E. coli* only during lactose utilization, and soon the study of lactose metabolism and beta-galactosidase became a site for the intersection of bacterial genetics and bacterial metabolism. The de novo synthesis of beta-galactosidase was induced by the presence of lactose as well as other substrates of this enzyme, and mutants defective in this process of induction, as well as the complementary sort of mutants that were always synthesizing beta-galactosidase, even when not needed, were found. The genetic dissection of lactose metabolism, as is well known, led Monod and his colleagues to propose the operon concept of gene regulation for which they were awarded the Nobel Prize in Physiology or Medicine in 1965. The parallel work on lysogeny, especially in Lwoff's group, soon established that the lysogenic state of a temperate phage and the induction of phage production from a lysogen was analogous to the induction of beta-galactosidase in the lactose operon. The prophage is, in fact, a set of genes under the negative control of a "repressor" molecule that, when inactivated, leads to the start of phage gene expression and eventually full phage reproduction. Phage mutants that could not form lysogens (clear plaque mutants, for example) were defective in the repression mechanism. The power of phage genetics joined with the biochemical under-

standing of bacterial gene regulation to finally explain the
mysteries of lysogeny that had persisted since the debates of
d'Herelle and Bordet.[9]

This rapprochement of the 1960s meant that the new-
generation phage researchers were facing adjustments and revisions in their methods, goals, and basic commitments. Target theory was all but forgotten. Postwar biochemistry brought
new tools for looking inside the cell, so the black-box mentality
was hard to retain. The genetics of the bacterial host was now
on firm ground. Structural biology was finally coming of age,
and key macromolecules such as DNA and a few proteins were
starting to yield to the x-ray crystallographers' model-build-
ing approaches. Electron microscopy was becoming available
to everyone, and expensive instrumentation was no longer restricted to a few Rockefeller-supported programs. Most importantly, theory was out and lab experiments were in. As one
wistful senior scientist said, "I thought that cleverness would
count, but now it is all just biochemistry."[10]

Although molecular biology claims phage biology as one
of its successful ancestors, the eclipse of the APG did not un-
dermine its future. Former phage biologists, some who were
members of the APG and some who were not, took their most
fundamental scientific commitments along with them as they
embarked on new endeavors: rejection of vitalism, a biological
materialism based on the physics and chemistry of molecules,
and the principle that genetics is a central tool and a basic epistemological fact of life.[11]

In a letter to Max Perutz (1914–2002) written in 1963,
Sydney Brenner defended his abandonment of phage in favor
of neurobiology:

It is now widely realized that nearly all the "classical" problems of molecular biology have either been

solved or will be solved in the next decade. The
entry of large numbers of American and other biochemists into the field will ensure that all the chemical details of replication and transcription will be
elucidated. Because of this, I have long felt that the
future of molecular biology lies in the extension of
research to other fields of biology, notably development and the nervous system. This is not an original thought because, as you well know, many other
molecular biologists are thinking in the same way.
The great difficulty about these fields is that the nature of the problem has not yet been clearly defined,
and hence the right experimental approach is not
known. There is a lot of talk about control mechanisms, and very little more than that.[12]

Molecular biology has not been without its dark moments, however. On occasion, perhaps in unguarded moments
or moods of melancholy, even its most successful practitioners
have expressed their existential doubts about the "program."
Brenner once told me what he saw as the dark fear of what he
called "the aldolase catastrophe": dread that the reductionist
approach to understanding life might end up amounting to a
vast collection of molecules, each as exciting as aldolase.[13] Delbrück, too, reflected on his disappointment at the mundane answer to the problem of how like begets like, the central question
of heredity. He noted that the finding that the genetic code is
copied by a replication mechanism based on complementary
base pairing—this mysterious aspect of life that had attracted
him to genetics in search of deep principles of nature—had
been reduced to the dull matter of stereochemistry of a few
small organic molecules. But maybe William of Ockham would
be pleased.

14

The American Phage Group as a Model of Discipline Formation

Molecular biology is now generally recognized as a field of science, an identifiable discipline with academic departments, degrees, and journals. Many biologists now distinguish molecular from organismal, or sometimes ecology and evolutionary, subdivisions of their interests, seeing distinctive research frameworks for these subdivisions, different enough to legitimately be considered disciplines. Even if one grants (which I do not) that "all of biology is now molecular," during the period between 1940 and 1960 when the APG was developing and thriving, "molecular biology" was not hegemonic as it is today.

The term "molecular biology" was first used by Warren Weaver in his annual report of 1937 for the Rockefeller Foundation, as he sought new directions for the foundation's biological research agenda. By 1959, the staid Cambridge University Press had launched the *Journal of Molecular Biology*. Sometime during these two decades, molecular biology as a discipline seems to have taken shape. By many accounts, a (or the) key component in the "origin story" of molecular biology was the success of research on bacteriophages and microbial genetics.[1] The earlier chapters in this book have described some of this research activity, carried out by a community of scientists later dubbed "the American Phage Group" by one of its prominent members, Gunther Stent. But just how do these two aspects of the story come together? How can the activities of the APG be seen as part of the disciplinary formation story of molecular biology? To make this connection clear, a more formal analysis seems needed. What, more precisely, constitutes discipline formation? What part did the APG play in the formation process?

Discipline formation is a fascinating and important topic in understanding science as a human endeavor, but it has been surprisingly understudied. New disciplines, new intellectual arrangements, and new social organizations of science are a constant feature in the history of science, and yet—except for crucial studies on topics such as the formation of scientific societies, religious influences on science, and "landmark experiments"—relatively little analysis has been given to the structure of discipline formation. Psychology as an outgrowth of nineteenth-century philosophy and physiology, and computer science as some multiparent offspring of mathematics, philosophy, electrical engineering, and cryptography are two familiar examples of new disciplines arising from old ones, reconfigured in response to contingent events, discoveries, and needs.[2] In the case of molecular biology, we have the well-known Delbrück festschrift, published in 1966 with the promising title *Phage and the Origins of Molecular Biology*, which, sadly, says little about the actual links between the two concepts in its title. The analysis of the APG published in 1972 by the sociologist Nicholas Mullins provides a starting point by tracing the evolution of the APG (relying on the Delbrück festschrift as his primary data source), using the related concepts of mentorship and joint publication as the keys to his analysis. Very recently, Manfred Laubichler and his colleagues have used data-mining approaches to extend the early work of Derek de Solla Price on citation analysis and the concept of the "invisible college" to analyze the evolution of the related fields of evolutionary and developmental biology (sometimes called "evo-devo").[3]

Thomas Kuhn's work on scientific revolutions lays out principles that suggest ways to conceive of this subject, and these have been developed by the philosopher Barbara Von Eckardt in her study of the formation of cognitive science as a distinct discipline. She describes her analysis as the "framework model" and, following Kuhn, she bases it on the fundamental notion of "shared commitments" by people engaged in related scientific activities. This "framework of shared commitments" provides, I think, a useful way to see how the APG has contributed in a central way to the origin of molecular biology. The shared assumptions and beliefs of a research community can exist even in the early stages of its development. These shared commitments help to define who is admitted as a member of the community. In her study of molecular evolution as a discipline, Edna Suarez-Diaz notes: "it is a question of *communication* in two or more dimensions . . . of devising concepts that are shared and collectively modified to address diverse epistemic goals." Concept formation, concept communication, concept sharing, and concept revision—all notions that we have seen play out in the life of the APG.

A framework of shared commitments can, and perhaps must, be transdisciplinary at the start of a new discipline. In the case of the APG, certain transdisciplinary ideas and assumptions can easily be identified. The stability as well as the evolution of these underlying principles will be the subject of this chapter. In this way, we can more clearly appreciate the convergence and development of a transdisciplinary commu-

nity of scientists into a well-defined, authoritative, and successful component of the new discipline of molecular biology.[4]

Von Eckardt suggests that one can identify both revisable
and unrevisable assumptions within a research framework. Stability resides in the unrevisable commitments. For the APG
these seem to include the following: (1) bacteriophages are viruses; (2) phages exhibit classical heredity behavior; (3) physical and chemical principles will be useful in understanding
heredity and reproduction. Of course, there were even more
basic commitments, common to most twentieth-century scientists, such as some sort of materialism, belief in a unitary
origin of life, and some version of Darwin's organic evolution,
together with a kind of linear reductionism where simple analysis of the parts of a problem could lead to understanding of
the whole based on a summation of the parts. One might consider these tenets as "background commitments," so commonly
held that they did not distinguish or even really concern their
adherents, a bit like the axioms of Euclidean geometry as differentiated from its postulates.

From their first foray into phage research, the founders
of the APG, Delbrück, Luria, and Hershey, were convinced of
(or at least adopted) these core commitments. In Delbrück's
first paper with Emory Ellis, the authors stated in the opening
sentence that they view bacteriophages as "viruses."[5] In the
same paper, they conclude that d'Herelle's view that phages
are viruses of bacteria is correct: "With our phage, our experiments confirm in the main the picture proposed by d'Herelle,
according to which a phage particle grows in the following
way: it becomes attached to a susceptible bacterium, multiplies
upon or within it up to a critical time, when the newly formed
phage particles are dispersed into the solution." Luria, in his
work with Holweck in France before coming to the United

States in 1940, clearly adopted this same viewpoint, necessary to his radiobiological target theory approach to measuring the sizes of phages (which assumed a particulate nature with a single lethal target per biologically active unit). Both Delbrück and Luria—like Delbrück's mentor in phage work, Ellis—were students of chemistry and physics, and less so of biology and medicine. Their education and experience pointed them toward explanations and understanding of biology rooted in these disciplines.

In 1944 Delbrück gave a series of lectures at Vanderbilt's School of Medicine on "Problems of Modern Biology in Relation to Atomic Physics," circulated privately to the grantees of the American Cancer Society. In these lectures he talked to biologists about the ways that chemistry and physics could and could *not* be applied to biology. It may seem surprising to us that although Delbrück arrived at Vanderbilt University in 1942, a time when American physicists were thinking about atom bombs, the subject of quantum mechanics was first taught there by Delbrück in 1947. Delbrück's 1944 lectures cast biological problems in the same mold as the problems of atomic physics: the dual problem of the parts and the whole, and the distinction between the process of observation and the process that was being observed. In physics, this latter problem was called "complementarity" by Delbrück's mentor, Niels Bohr, and "indeterminacy" (or uncertainty) by his only slightly older former colleague Werner Heisenberg. One of the core principles of Delbrück's biology (also held, as it turned out, by most of the APG) was that simplicity and reductionism were essential to retain, even as one faced the dilemma inherent in their work: having to tear up living things to study their basic properties yet in the process destroying the very object one desires to study—that is, life itself.

In his first lecture, Delbrück set the stage for the series but also provided his version of the shared intellectual commitments of the new field he hoped to establish:

> The most notable of these [fields of biology] is genetics, which in its most pure form operates with "hereditary factors" and "phenotypic characters" in a perfectly logical system, without ever having to bother with processes by which the "characters" originate from the "factors." The root of this science lies in the existence of natural units of observation, the individual living organisms which in genetics play much the same role as the atom and molecules in chemistry.
>
> However, though the geneticists may be content to stop here, chemists and physicists will not. They will want to go beyond the descriptions offered by pure genetics, as a result of which we have today "physiological genetics" as well, by which genes influence the phenotype of individuals. Physicists and chemists will also want to go beyond the geneticists by inquiring into the mechanism of the reproduction of the gene and into the peculiar phenomenon that if a gene is changed it will reproduce true to its changed form, a kind of covariant reproduction. Is this process simply a trick of organic chemistry which the chemists have not yet run across in their test tubes or is it a phenomenon of an entirely different nature, depending on the complex "organization" of the living cell? The answer to this question is at present totally unknown. The old-line mechanists, of course, would consider it sacrilegious thought that the reproduction of genes might

not be a purely chemical process, but the modern physicist may well be inclined to take a more liberal view.

Let us have a look at the modern physicist who seems so willing to sell what seemed to be his birthright.[6]

Chemistry and physics would gain a hold over biology by virtue of two great generalizations. Chemistry showed that living material is made up of the same elements as the materials of the inanimate world, and physics showed that conservation of energy and certain other physical "laws" are valid for processes occurring in living material just as they are for all processes in the inanimate world. Delbrück conceded that there were limits on scientific observation: "the distinction between the observing tool and the object of observation which we have to make at some arbitrary point necessitates a certain latitude in our description of the object. This situation, the analysis of which is due chiefly to Bohr and Heisenberg, has been termed 'the principle of indeterminacy' by Heisenberg, and 'the principle of complementariness' by Bohr."[7]

It is important to recall some of the dominant notions of virology at the time to understand how strange, and perhaps even revolutionary, these ideas about how to study phages must have sounded to most biologists. Virology was situated firmly in the field of disease pathology. Both animals and plants were known to be afflicted by infections that could be transmitted by solutions that passed through filters (hence "filterable") that removed microscopically visible microbes: bacteria, protozoa, and fungi. Only recently had the qualifying adjective "filterable" been dropped from the usual references to "filterable viruses." Viruses were observed almost exclusively by their pathological effects and physiological disruptions of host organisms. Labo-

ratory animals such as rodents were observed for pathogno-
monic lesions, or more often simply survival. Plants, such as
tobacco, had to be injured with abrasives to allow entry of viruses that could only be assayed by the production of characteristic lesions. Cytological effects of such infections, often seen
as stained "inclusion bodies," were detected, but the relation of
these cytochemical observations to virus biology was unclear.
Quantitation of the "amount" of virus was uncertain at best,
with virus stocks often simply described as strong or weak.
Against this state of affairs that had evolved from classical natural history, the APG believed that chemical and physical principles that governed all matter could lead to a better understanding of some of the basic properties of all living things.

Historians and philosophers of science have recognized
that scientists choose objects of study that will, when manipu-
lated, produce "facts" that can be interpreted in ways that allow
their theories and beliefs to make sense. Sometimes the short-
hand "the right tool for the job" is suggested to describe such
choices. But this notion elides the fact that what "the job" is
must be determined before the tool can be selected. In the case
of the APG, "the job" was one of the core assumptions of the
research framework. It was nothing less than the biblical question of "how like begets like." As noted earlier, current scientific discourse speaks of "replication" of genes and organisms,
but for the APG, certainly in the beginning at least, the discussion focused on "reproduction," the process of faithful and
highly accurate production of offspring (together with the cor-
ollary problem of rare variation—that is, mutation—essential
to organic evolution). Understanding reproduction was seen
as the gold ring of their scientific program.

Delbrück's reference to "conservation of energy" signaled
another key commitment within the framework shared by the
new community of phage workers. The basic question of "what

is life?" was to be guided by the great nineteenth-century physical edifice of thermodynamics. To the physicist, energy and its behavior transcended all else. Hydrogen atoms, genes, and human thought were all subject to the laws of thermodynamics. Physicists embraced a fully materialist view of the living state and sought theories and experiments to make clear just how thermodynamics and energy could explain the essence of life. *Biological theories were to be tested against the laws of physics and chemistry.* In this program, they sided with the burgeoning community of biological chemists who were making important advances in describing the metabolic activities of living systems in terms of the various chemical reactions that could be detected both in living cells and in material that was once part of a living cell. Might not life be just a quasi-stable complex system of multiple molecular reactions governed by a temporarily favorable configuration of packets of free energy? Likewise, the mysterious process of how like begets like must be governed by the same thermodynamic laws; physics would be the way to study reproduction if only one could find the right system to lay bare these laws in action.

Another shared commitment—not so obvious, however—was an anti-reductionist belief that led away from hydrogen atoms and required a truly living organism to study. But not so anti-reductionist as to get bogged down in descriptive study of sea urchin embryos or even fruit fly metamorphosis. Instead, it appeared to the APG that bacteriophages were almost like the hydrogen atom of biology. Seemingly simple yet endowed with the apparent qualities of all living things, phage was the object (or "gadget" as suggested by Lily Kay) that promised to become a biology "fact-making machine" that might unlock some of the basic secrets of life.[8]

The third unrevisable shared commitment of the APG related to the problem of stability of the organism undergoing reproduction. Heredity should be "faithful." Only rarely do variants arise. Indeed, in the first half of the twentieth century, the problem of mutation was a central concern in biology, important because it was key to the "modern synthesis" in which Darwin's ideas of variation and natural selection were taken together with genetic ideas of Mendel, Weismann, and Morgan to provide a more explicit description of organic evolution. Mutation, however, was also on the minds of the physicists who had been struggling to understand rare but discrete transitions in energy states of atoms. Were biological mutations just a manifestation of the same sort of physical changes being described by quantum mechanics? It seems that was the first idea of both Delbrück and Schrödinger. The physicist Schrödinger wondered: "How are we to understand that it [the trait of the Hapsburg lip] has remained unperturbed by the disordering tendency of the heat motion for centuries? . . . In this case the answer is supplied by quantum theory. In light of present knowledge, the mechanism of heredity is closely related to, nay, founded on, the very basis of quantum theory. . . . The physical aspect leaves no other possibility to account for its [the gene's] permanence."[9]

The founders of the APG saw viruses, and especially bacteriophages, as the right unit of analysis to get at the physical nature of the gene and the understanding of its remarkable stability yet ability to change abruptly and discontinuously to a new form; the analogy to quantum transitions was too tempting to ignore. The geneticist H. J. Muller presciently observed in 1922 that a bacteriophage particle might be a "naked gene."[10] Like atoms, phage in the early days had few observable properties on which to base theoretical models, but those that did exist (plaque morphology, host range, and antigenic determinants) were well-behaved indicators of biological heredity.

It is important to note, too, the *absence* of certain beliefs that would only later be incorporated into the research framework of the APG. Most notable was the lack of any strong consensus as to the chemical nature of the gene or the detailed chemical structure of the phage particle itself. Initially, the gene was treated as a "black box" with certain observable properties and behaviors, inputs and outputs, if you will. These did not extend to traditional chemical qualities. When Alfred Hershey and Martha Chase did their famous "blender experiment," Hershey was, according to contemporary accounts, expecting that the protein component of the phage would enter the cell and carry out the genetic function of the phage. Likewise, although DNA was considered important—after all, it comprised roughly half of the mass of the phage particles—the double helical model of DNA proposed by Watson and Crick in 1953 was not seen as unproblematic.[11]

As a research community develops, it may, without relinquishing its core framework, add to or revise its shared commitments by incorporating ideas, material, and theories from other communities. In the case of the APG, two early additions were the admission of some biochemical ideas and the expansion of the initial restriction of attention to just the T-phages to include the study of λ phage and the phenomenon of lysogeny.

From the start, the more usual studies of macromolecular chemistry were applied to phages but were not part of the core commitments. A chemical description of viruses, initially started by Max Schlesinger and culminating in the crystallization and chemical analyses of TMV by John Desmond Bernal, Wendell M. Stanley, Fred Bawden (1908–1972), and Norman W. Pirie (1907–1997) suggested that knowing something about the chemical makeup of viruses might be helpful to the more physical approaches of the initial APG scientists. Seymour Cohen, a student of Wendell Stanley, and the group led by Earl Evans, which included Frank Putnam, Lloyd Kozloff, and Roy P. Mackal (1925–2013), gave consistent voice to the relevance of biochemical studies of phages to the early APG. While these researchers seemed to be very visible in the APG community, none of them considered themselves as true members of the APG.

Because Delbrück had decreed that only the seven fully lytic T-phages were legitimate experimental objects, phages that could exist in latent form in the host lysogenic bacteria were excluded from the initial version of shared commitments of the APG. Indeed, Delbrück was suspicious that such phages and the so-called lysogenic state might be some sort of experimental artifact. With the isolation of phage λ in 1952 by Esther Lederberg, however, so-called temperate phages started to become integrated into the thinking of the APG. For one thing, mutants of λ that were strictly lytic phages suggested that temperate phages were not very different from the canonical T-phages. These mutants could grow in lysogenic strains, thus overcoming the "immunity" of lysogenic bacteria to infection by their cognate phages. These mutants could not enter the lysogenic state and thus behaved like the T-phages. Soon λ phage became an accepted object for the APG, and the study of the process of lysogenization, "immunity," and induction of phage production from lysogenic bacteria greatly expanded the theoretical and experimental purview of the phage community. The discovery that the mutation that suppressed lysogenization and gave rise to clear rather than turbid plaques was in a gene that repressed most of the gene expression of the other genes of λ provided an important overlap with the parallel work on bacterial gene expression in the lactose fermentation pathway, studies that led Jacob, Monod, and their collaborators at the Pasteur Institute to develop the generality of the operon model of gene regulation.[12]

One milestone in the growth of a scientific field is the integration and interpretation of its core commitments into existing, accepted scientific beliefs of related disciplines. By the late 1950s and early 1960s, both biochemistry and microbial genetics were fields that accepted and were enriched by results from phage biology. Conversely, the young field of phage biology saw its expanding, shared core commitments incorporating new results from both biochemistry and genetics. This syncretism contributed greatly to the foundational solidity of the nascent discipline that was forming around phage biology, a research community that would soon become a key progenitor in the birth of molecular biology. A small group of pioneer phage workers, led by an even smaller nucleus of charismatic founders, effectively transplanted their shared intellectual and social commitments into a central domain of biology—that is, genetics—and, in the process, the American Phage Group transformed and modernized the entire discipline of biology.[13]

Notes

1
Life, Genes, and Phages

1. Robert Olby, *The Molecular Revolution in Biology* (Abingdon-on-Thames, U.K.: Routledge, 2020).

2. Hermann J. Muller, "Physics in the Attack on the Fundamental Problems of Genetics," *Scientific Monthly* 44, no. 3 (1937): 210–214.

3. Richard B. Goldschmidt, *Theoretical Genetics* (Berkeley: University of California Press, 1955), 95.

4. Philip Kitcher, "Genes," *British Journal for the Philosophy of Science* 33, no. 4 (1982): 337–359.

5. Samuel J. Keyser, George A. Miller, and Edward Walker, "Cognitive Science, 1978: Report of the State-of-the-Art Committee to the Advisors of the Alfred P. Sloan Foundation."

6. Robert C. Olby, *The Path to the Double Helix* (Seattle: University of Washington Press, 1974); Horace Freeland Judson, *The Eighth Day of Creation: Makers of the Revolution in Biology* (New York: Simon and Schuster, 1979). For a more comprehensive and focused study of the origins of molecular biology, see Lily E. Kay, *The Molecular Vision of Life: Caltech, the Rockefeller Foundation, and the Rise of the New Biology* (New York: Oxford University Press, 1993), and Michel Morange, *A History of Molecular Biology,* translated by Matthew Cobb (Cambridge: Harvard University Press, 1998); see also Joseph S. Fruton, *Proteins, Enzymes, Genes: The Interplay of Chemistry and Biology* (New Haven: Yale University Press, 1999); Ulf Lagerkvist, *DNA Pioneers and Their Legacy* (New Haven: Yale University Press, 1998); John Cairns, Gunther S. Stent, and James D. Watson, eds., *Phage and the Origins of Molecular Biology* (Cold Spring Harbor, N.Y.: Cold Spring Harbor

Laboratory Press, 1966), informally known as *PATOOMB* (unless otherwise specified, all citations are to this 1966 edition); Salvador E. Luria, *A Slot Machine, a Broken Test Tube: An Autobiography* (New York: Harper & Row, 1984); Ernst Peter Fischer and Carol Lipson, *Thinking About Science: Max Delbrück and the Origins of Molecular Biology* (New York: Norton, 1988); Gunther S. Stent, *Nazis, Women, and Molecular Biology: Memoirs of a Lucky Self-Hater* (Kensington, Calif.: Briones Books, 1998).

7. Cairns, Stent, and Watson, *Phage and the Origins;* historians: Donald Fleming and Bernard Bailyn, eds., *The Intellectual Migration: Europe and America, 1930–1960* (Cambridge: Belknap Press of Harvard University Press, 1969); sociologists: Nicholas C. Mullins, "The Development of a Scientific Specialty: The Phage Group and the Origins of Molecular Biology," *Minerva* 10, no.1 (1972): 51–82; scientists: Robin Holliday, "Physics and the Origins of Molecular Biology," *Journal of Genetics* 85, no. 2 (2006): 93–97; Max Perutz, "Origins of Molecular-Biology," *New Scientist* 85, no. 1192 (1980): 326–329; John Cairns, Gunther S. Stent, and James D. Watson, eds., *Phage and the Origins of Molecular Biology,* Centennial edition (Cold Spring Harbor, N.Y.: Cold Spring Harbor Laboratory Press, 2007).

2

Phage Before the War (1917–1940)

1. William C. Summers, "On the Origins of the Science in *Arrowsmith:* Paul De Kruif, Félix d'Herelle, and Phage," *Journal of the History of Medicine and Allied Sciences* 46 (1991): 315–332; William C. Summers, *Félix d'Herelle and the Origin of Molecular Biology* (New Haven: Yale University Press, 1999).

2. Hansjürgen Raettig, *Bakteriophagie, 1917 bis 1956,* 2 vols. (Stuttgart: G. Fischer, 1958).

3. William Whiteman Carlton Topley and Graham Selby Wilson, "The Twort-d'Herelle Phenomenon," in *The Principles of Bacteriology and Immunity,* vol. 1 (New York: Williams and Wilkins, 1929), 224–233; Hans Zinsser and Stanhope Bayne-Jones, "Bacteriophage," in *Textbook of Bacteriology,* 8th ed. (New York: Appleton Century, 1937), 837–842.

4. William C. Summers, "How Bacteriophage Came to Be Used by the Phage Group," *Journal of the History of Biology* 26 (1993): 255–267.

5. The potential for phage therapy galvanized the enthusiasm of young physicians especially. In the United States Martha Wollstein, Ann Gayler Kuttner, Earl B. McKinley, Wilbur C. Davidson, and Edwin W. Schultz— who would all become distinguished academicians—contributed to the literature and exploration of phage therapy. In Europe and Asia as well, many

young physicians wrote their doctoral theses and conducted initial clinical trials on phage.

6. Félix d'Herelle, "Exogenous Immunity: Bacteriophagy *in vivo*," in *Immunity in Natural Infectious Disease*, translated by George H. Smith (Baltimore: Williams and Wilkins, 1924), 271–307; André Gratia, "Studies on the d'Herelle Phenomenon," *Journal of Experimental Medicine* 34, no. 1 (1921): 115–126; "Jacques Jacob Bronfenbrenner (1883–1953)," Bernard Becker Medical Library at Washington University School of Medicine website, accessed 25 May 2021, http://beckerexhibits.wustl.edu/mig/bios/bronfenbrenner.html.

7. Ton van Helvoort, "Bacteriological and Physiological Research Styles in the Early Controversy on the Nature of the Bacteriophage Phenomenon," *Medical History* 36, no. 3 (1992): 243–270.

8. Elie Wollman, interview by author, Paris, 7 June 1989.

9. Marie Curie, "Sur l'étude des courbes de probabilité relatives à l'action des rayons X sur les bacilles," in *Comptes rendus hebdomadaires des séances de l'Académie des science* 188 (1929): 202–204; Fernand Holweck, Salvatore [Salvador] Luria, and Eugène Wollman, "Recherches sur le mode d'action des radiations sur les bacteriophages," in *Comptes rendus hebdomadaires des séances de l'Académie des science* 210 (1940): 639–642.

10. Luria, *A Slot Machine*, 19–20; Geo Rita, "Su di un metodo pratico per l'isolamento del batteriofago," *Annali di igiene* 49 (1939): 661–664.

11. Obituary, "Max Schlesinger, M.D. Budapest," *Lancet* 229 (1937): 413. Some sources incorrectly give his first name as Martin; see also Heinrich Bechhold and Max Schlesinger, "Zentrifuge und Filter zur Bestimmung der absoluten Größe von subvisiblen Erregern," *Zeitschrift für Hygiene und Infektionskrankheiten* 112, no. 4 (1931): 668–679; Max Schlesinger, "Reindarstellung eines Bakteriophagen in mit freiem Auge sichtbaren Mengen," *Biochemische Zeitschrift* 264 (1933): 6–12; Max Schlesinger, "Beobachtung und Zählung von Bakteriophagenteilchen im Dunkelfeld. Die Form der Teilchen," *Zeitschrift für Hygiene und Infektionskrankheiten* 115, no. 4 (1933): 774–780; Max Schlesinger, "Zur frage der chemischen zusammensetzung des bakteriophagen," *Biochemische Zeitschrift* 273 (1934): 306–311; Max Schlesinger, "The Feulgen Reaction of the Bacteriophage Substance," *Nature* 138 (1936): 508–509.

12. Nikolai Boulgakov (Bulgakov) was the younger brother of the Soviet novelist Mikhail Bulgakov; see Paul Bonét-Maury and Nikolai Bulgakov, "Recherches sur la taille et la structure du Bacteriophage φX174: Action des rayons alpha du radon," *Comptes rendus des séances de la Société de biologie et de ses filiales* 138 (1944): 499; Paul Bonét-Maury, "L'irradiation des virus," in *Proceedings of the International Conference on Radiobiology* (1954): 75–78; Nikolai Bulgakov and Paul Bonét-Maury, "Recherches sur la taille et la struc-

ture du Bacteriophage φX174: Méthode de titrage," *Comptes rendus des séances de la Société de biologie et de ses filiales* 138 (1944): 497; also Zdravko Lacković and Karlo Toljan, "Vladimir Sertić: Forgotten Pioneer of Virology and Bacteriophage Therapy," *Notes and Records: The Royal Society Journal of the History of Science* 74, no. 4 (2020): 567–578.

13. Frank Macfarlane Burnet and Dora Lush, "Induced Lysogenicity and Mutation of Bacteriophage within Lysogenic Bacteria," *Australian Journal of Experimental Biology and Medical Science* 14 (1936): 27–38; see also Neeraja Sankaran, "Frank Macfarlane Burnet and the Nature of the Bacteriophage, 1924–1937" (Ph.D. dissertation, Yale University, 2006).

3
From Physics to Biology

1. Garland E. Allen, "Opposition to the Mendelian-Chromosome Theory: The Physiological and Developmental Genetics of Richard Goldschmidt," *Journal of the History of Biology* 7, no. 1 (1974): 49–92.

2. Alexander Gurwitsch, *Mitogenetic Emission* (Moscow: Gos. Med. Izdat., 1932).

3. Luis A. Campos, *Radium and the Secret of Life* (Chicago: University of Chicago Press, 2015).

4. Hermann J. Muller, "The Production of Mutations by X-rays," *Proceedings of the National Academy of Sciences of the United States of America* 14, no. 9 (1928): 714–726.

5. Nikolay V. Timoféef-Ressovsky, Karl G. Zimmer, and Max Delbrück," Über die Natur der Genmutation und der Genstruktur," *Nachrichten Gesellschaft der Wissenschaften zu Göttingen, math-phys* Kl. (1935): 189–245; Erwin Schrödinger, *What Is Life and What Is Matter* (Cambridge: Cambridge University Press, 1967), 61.

6. Cairns, Stent, and Watson, *Phage and the Origins,* 37.

7. Gunther S. Stent, *Molecular Biology of Bacterial Viruses* (San Francisco: W. H. Freeman, 1963), 18.

8. Douglas Edward Lea, *Actions of Radiations on Living Cells* (Cambridge: Cambridge University Press, 1946); Richard B. Setlow and Ernest C. Pollard, *Molecular Biophysics* (Boston: Addison-Wesley, 1962).

9. John Heilbron, "The Scattering of α and β Particles and Rutherford's Atom," *Archives of the History of the Exact Sciences* 4 (1968): 247–307; John J. Thomson, "On the Number of Corpuscles in an Atom," *Philosophical Magazine* 11 (1906): 769–781.

10. Ernest Rutherford, "Collision of α-Particles with Light Atoms, I–IV,"

Philosophical Magazine 37 (1919): 537–587; see also James Chadwick and Etienne S. Biéler, "The Collision of α-Particles with Hydrogen Nuclei," *Philosophical Magazine* 42 (1922): 923–940.

11. Charles T. R. Wilson, "On an Expansion Apparatus for Making Visible the Tracks of Ionizing Particles in Gasses and Results Obtained from Its Use," *Proceedings of the Royal Society of London. Series A* 87 (1912): 277–292.

12. Thomas S. P. Strangeways and H. E. H. Oakley, "The Immediate Changes Observed in Tissue Cells After Exposure to Soft X Rays While Growing *in vitro*," *Proceedings of the Royal Society of London. Series B* 95 (1923): 373–381. H. E. H. Oakley took up the practice of radiology in Durban, South Africa, in February 1923. James A. Crowther, "Some Considerations Relative to the Action of X Rays on Tissue Cells," *Proceedings of the Royal Society of London. Series B* 96 (1924): 207–211. By a probability law, Crowther means the Poisson distribution. This distribution plots the probability of a given number of events (hits) occurring in a fixed unit of time or space (the target) assuming these events occur with a known constant mean rate and each event is independent of the prior event.

13. Friedrich Dessauer, "Über einige Wirkungen von Strahlen. I," *Zeitschrift für Physik* 12 (1923): 38–47; Marietta Blau and Kamillo Alternburger, "Über einige Wirkungen von Strahlen. II," *Zeitschrift für Physik* 12 (1923): 315–329; Friedrich Dessauer, "Über einige Wirkungen von Strahlen. IV," *Zeitschrift für Physik* 20 (1923): 288–298; Friedrich Dessauer, "Zur Erklärung de biologische Strahlenwirkungen," *Strahlentherapie* 16 (1924): 208–221. See Hubert Goenner, "Albert Einstein and Friedrich Dessauer: Political Views and Political Practice," *Physics in Perspective* 5, no. 1 (2003): 21–66; also, Peter L. Galison, "Marietta Blau: Between Nazis and Nuclei," *Physics Today* 50, no. 11 (1997): 42–48; see also Dessauer, "Wirkungen von Strahlen I," 40.

14. Equation (3) from Dessauer, "Wirkungen von Strahlen I," 44.

15. Dessauer, "Wirkungen von Strahlen I," 45.

16. Emil Warburg, "Über Plancks Verdienste um die Experimentalphysik," *Naturwissenschaften* 6, no. 17 (1918): 202–203; Dessauer, "Wirkungen von Strahlen I," 46–47.

17. Dessauer, "Wirkungen von Strahlen IV," 293.

18. Edward U. Condon and H. M. Terrill, "Quantum Phenomena in the Biological Action of X Rays," *Journal of Cancer Research* 11 (1927): 324–333.

19. Douglas E. Lea, R. B. Haines, and Charles Alfred Coulson, "The Mechanism of the Bactericidal Action of Radioactive Radiations. I. Theoretical," *Proceedings of the Royal Society of London. Series B* 120, no. 816 (1936): 47–76. The collection of D. E. Lea papers and notebooks, 1931–1943, is held in the United States by the Special Collections Library of the University of Tennessee.

20. Sinclair Lewis, *Arrowsmith* (New York: Harcourt, Brace, 1925), 406.
21. James Dewey Watson, "The Properties of X ray-inactivated Bacteriophage. I. Inactivation by Direct Effect," *Journal of Bacteriology* 60 (1950): 697–718; and "II. Inactivation by Indirect Effects," *Journal of Bacteriology* 63 (1952): 473–485.

4

Heterogeneity: Physicists Doing Biology

1. See, for examples: Evelyn Fox Keller, "Physics and the Emergence of Molecular Biology: A History of Cognitive and Political Synergy," *Journal of the History of Biology* 23 (1990): 389–409; Fischer and Lipson, *Thinking About Science*, 46–69. Delbrück's career as a physicist got off to an inauspicious start when he failed his doctoral exam. His research contributions in physics seem to have been solid but undistinguished. Michel Morange, *Histoire de la biologie moléculaire* (Paris: La Decouverte, 1994), 89–104.
2. Almost without exception, biology and chemistry textbooks (even to the present day) include an introductory chapter that makes reference to Wöhler's famous synthesis of urea (a compound thought to be only the product of "life") in 1828, as a stake in the heart of the old "vitalism" of the past. See Peter J. Ramberg, "The Death of Vitalism and the Birth of Organic Chemistry: Wöhler's Urea Synthesis and the Disciplinary Identity of Organic Chemistry," *Ambix* 47, no. 3 (2000): 170–195.
3. The next generation of philosophers of science in Vienna would aggressively explore this "unity of science" idea. Rudolf Carnap (1891–1970) and Otto Neurath (1882–1945) in the 1930s engaged both philosophers and physicists in detailed expositions of two different approaches to the unity of science program. Biology and other "immature sciences" would mature when they became more like physics, and it was the physicists who could guide this maturation.
4. Pierre Serge Choumoff, "Fernand Holweck (1890–1941)," in *Vacuum Science and Technology: Pioneers of the 20th Century*, ed. Paul A. Redhead (New York: American Institute of Physics Press, 1994), 59–67.
5. Luria, *A Slot Machine*, 22–23.
6. Niels Bohr, "Light and Life," *Nature* 131 (1933): 457–459; Gunther S. Stent, "Light and Life: Niels Bohr's Legacy to Contemporary Biology," *Genome* 31, no. 1 (1989): 11–15.
7. Hans Gaffron, "Dinner Talk, University of Chicago," n.d., p. 10. Harold C. Urey Papers, University of California San Diego Library, Special Collection. Mss. 0044, Box 35, folder 1.

8. For recent work on Jordan see Richard H. Beyler, "Targeting the Organism: The Scientific and Cultural Context of Pascual Jordan's Quantum Biology, 1932–1947," *Isis* 87 (1996): 248–273; also M. Norton Wise, "Pascual Jordan: Quantum Mechanics, Psychology, National Socialism," in *Science, Technology, and National Socialism,* ed. Monika Rennenberg and Mark Walker (Cambridge: Cambridge University Press, 1994), 224–254; Finn Aaserud, *Redirecting Science: Niels Bohr, Philanthropy, and the Rise of Nuclear Physics* (Cambridge: Cambridge University Press, 1990), 83–90; Pascual Jordan, "Die Quantenmechanik und die Grundprobleme der Biologie und Psychologie," *Naturwissenschflen* 20 (1932): 815–821.

9. Quoted in Beyler, "Targeting the Organism," 260.

10. Juan A. Del Regato, "Friedrich Dessauer," *International Journal of Radiation Oncology: Biology, Physics* 4, no. 3 (1978): 325–332.

11. Paul Forman, "Weimar Culture, Causality, and Quantum Theory, 1918–1927: Adaptation by German Physicists and Mathematicians to a Hostile Intellectual Environment," *Historical Studies in the Physical Sciences* 3 (1971): 1–115.

12. Spencer Weart and Gertrude W. Szilard, eds., *Leo Szilard: His Version of the Facts: Selected Recollections and Correspondence* (Cambridge, Mass.: MIT Press, 1979), 17–18.

13. Leo Szilard, "On the Nature of the Aging Process," *Proceedings of the National Academy of Sciences of the United States of America* 45, no. 1 (1959): 30–45.

14. Ernest C. Pollard, *Radiation: Cells and People* (Lemont, Pa.: Woodburn Press, 1990), 27, 93–94.

15. Pollard, *Radiation,* 27–28.

16. Pollard, *Radiation,* 28; also, interview by author (1996) with Franklin Hutchinson, a member of Pollard's group at MIT and later his graduate student and colleague at Yale.

17. Douglas E. Lea, Kenneth M. Smith, Barbara Holmes, and Roy Markham, "Direct and Indirect Actions of Radiation on Viruses and Enzymes," *Parasitology* 36, no. 1–2 (1944): 110–118; quotation from Pollard, *Radiation,* 29.

18. Pollard, *Radiation,* 30.

19. Ernest C. Pollard, *The Physics of Viruses* (New York: Academic Press, 1953), v.

20. Setlow and Pollard, *Molecular Biophysics.*

21. Eduard Kellenberger, "History of Phage Research as Viewed by a European," *FEMS Microbiology Reviews* 17 (1995): 7–24; see also John R. Günter, ed., *History of Electron Microscopy in Switzerland* (Basel: Birkhäuser, 1990).

22. Timoféef-Ressovsky, Zimmer, and Delbrück, "Genmutation und der Genstruktur," 189–245. Although Auger, the discoverer of Auger electrons,

confined his research to quantum physics and cosmic ray physics, he was an interested and enthusiastic supporter of genetics. As director of higher education at the Sorbonne from 1945 to 1948, he created the first chair of genetics there and awarded it to Boris Ephrussi. Ervin Bauer was listed on the Woods Hole program, according to Demerec, but since Bauer was murdered by the Soviet NKVD in January 1938, Demerec apparently relied on preliminary information. Milislav Demerec, "Memorandum Regarding, Dr. Ugo Fano," 10 September 1940. Cold Spring Harbor Laboratory Archives: Carnegie Files, Fano folder. Timoféef was a "guest-associate" at the Cold Spring Harbor Laboratories for two to three months in the fall of 1932 after his participation in the famous Sixth International Congress on Genetics held in 1932 at Ithaca, New York. Letter from Demerec to Timoféef-Ressovsky, 30 January 1932, Milislav Demerec Papers, Mss. B.D394, American Philosophical Society Archives; Carnegie Institution of Washington, Annual Report, 1932–1933, 32.

23. Demerec, "Memorandum." Fano and Luria were not only childhood friends in Turin but also had mutual connections with Fermi in Rome, prior to their emigration to the United States.

24. Milislav Demerec and Ugo Fano, "Bacteriophage-resistant Mutants in *Escherichia coli*," *Genetics* 30 (1945): 119–136; letter from Demerec to Delbrück, 23 October 1945, Cold Spring Harbor Laboratory Archives: Carnegie Files, Delbrück folder; Charles W. Clark, "Obituary: Ugo Fano (1912–2001)," *Nature* 410, no. 6825 (2001): 164–165.

25. Anonymous, "Thinking Ahead with Leo Szilard," *International Science and Technology*, May (1962): 36.

26. Letter from Gunther Stent to William B. Treumann, 11 June 1947, Gunther Stent Papers, Carton 14, folder 53, Bancroft Library, University of California, Berkeley.

27. Letter from Gunther Stent to William Treumann, 12 May 1948, Gunther Stent Papers, Carton 14, folder 53, Bancroft Library, University of California, Berkeley.

28. Letter from Max Delbrück to Wendell Stanley, 29 June 1945, Wendell Stanley Papers, Carton 7, folder 145, Bancroft Library, University of California, Berkeley, permission courtesy of the Delbrück family; letter from Bjorn Sigurdsson to Wendell Stanley, 23 October 1945, Wendell Stanley Papers, Carton 10, folder 5, Bancroft Library, University of California, Berkeley.

29. Letter from George Gamow to Wendell Stanley, 24 February 1947, Wendell Stanley Papers, Carton 8, folder 77, Bancroft Library, University of California, Berkeley. By the spring of 1954, Gamow was noted to be devoting himself "one day a week to 'Biology.'" Letter from Gunther Stent to Max

Delbrück, 26 March 1954, Gunther Stent Papers, Carton 4, folder 20, Bancroft Library, University of California, Berkeley.

30. Nicolas Rasmussen, "The Mid-Century Biophysics Bubble: Hiroshima and the Biological Revolution in America, Revisited," *History of Science* 35, no. 3 (1997): 245–293.

31. Leo Koenigsberger, *Hermann von Helmholtz*, translated by Frances A. Welby (Oxford: Clarendon Press, 1906), 340.

32. David A. Grandy, *Leo Szilard: Science as a Mode of Being* (Lanham, Md: University Press of America, 1996), 104; Léon Rosenfeld, in *Niels Bohr: His Life and Work as Seen by His Friends and Colleagues*, ed. Stefan Rozental (Amsterdam: North-Holland Publishing Company, 1967), 130.

33. Kay, *Molecular Vision of Life*, chapter 8.

34. Ever since the nineteenth-century realization that there seemed to be "factors," "gemules," or some other theoretical entity by which visible characters of an organism were determined, biologists sought the causal and mechanistic link between these factors, eventually called the genotype by Sutton and Boveri in 1903, and their visibly expressed manifestations, called the phenotype; see also William Bateson, *Mendel's Principles of Heredity* (Cambridge: Cambridge University Press, 1909).

35. Bohr, "Light and Life."

36. Kay, *Molecular Vision of Life*.

37. This, in spite of Delbrück's interest in Bohr's famously vague suggestion of some sort of principle of biological complementarity, an analogy to a consequence of the then fashionable Copenhagen Interpretation of quantum mechanics. See Cairns, Stent, and Watson, *Phage and the Origins*, 20–22.

5
Heterogeneity: Chemists and Biologists

1. Lily E. Kay, "The Tools of the Discipline: Biochemists and Molecular Biologists," *Journal of the History of Biology* 29, No. 3 (1996): 447–450. See also Bernard S. Strauss, "A Physicist's Quest in Biology: Max Delbrück and 'Complementarity,'" *Genetics* 206 (2017): 641–650, and Max Delbrück, "A Physicist Looks at Biology," in Cairns, Stent, and Watson, *Phage and the Origins*, 22.

2. Max Delbrück, "A Physicist Looks at Biology," *Transactions of the Connecticut Academy of Arts and Sciences* 38 (1949): 173–190.

3. Gunther S. Stent, *Nazis, Women, and Molecular Biology*, 349.

4. A distinguished colleague of mine once quipped, upon entering his

undergraduate physics final exam: "All I have to remember is F=ma and I can derive the rest."

5. James D. Watson, "Growing Up in the Phage Group," in Cairns, Stent, and Watson, *Phage and the Origins*, 241.

6. For one view of this relationship see Robert E. Kohler, *From Medical Chemistry to Biochemistry: The Making of a Biomedical Discipline* (Cambridge: Cambridge University Press, 1982).

7. John W. Servos, *Physical Chemistry from Ostwald to Pauling: The Making of a Science in America* (Princeton: Princeton University Press, 1990); Gilbert Newton Lewis and Merle Randall, *Thermodynamics and the Free Energy of Chemical Substances* (New York: McGraw-Hill, 1923); Félix d'Herelle, "Bacteriophage, a Living Colloidal Micell," in *Colloid Chemistry, Theoretical and Applied*, vol. II, ed. J. Alexander (New York: Chemical Catalog, 1928), 535–541.

8. William T. Astbury, "X-ray Studies of the Structure of Compounds of Biological Interest," *Annual Review of Biochemistry* 8, no. 1 (1939): 113–133; Max S. Dunn, "Chemistry of Amino Acids and Proteins," *Annual Review of Biochemistry* 10, no. 1 (1941): 113; also Marjorie Senechal, *I Died for Beauty: Dorothy Wrinch and the Cultures of Science* (Oxford: Oxford University Press, 2012); and Joseph S. Fruton, "Energy-Rich Bonds and Enzymatic Peptide Synthesis," in *The Roots of Modern Biochemistry: Fritz Lipmann's Squiggle and Its Consequences*, ed. Horst Kleinkauf, Hans von Döhren, and Lothar Jaenicke (Berlin: Walter de Gruyter, 2011), 165–180.

9. Astbury, "X-ray Studies," 123.

10. For Stanley's bridge between organic and biochemistry of the 1930s, see Angela N. H. Creager, *The Life of a Virus: Tobacco Mosaic Virus as an Experimental Model, 1930–1965* (Chicago: University of Chicago Press, 2002); James Batcheller Sumner, "The Chemical Nature of Enzymes," *Journal of the Washington Academy of Sciences* 38, no. 4 (1948): 113–117. John H. Northrop, "The Chemistry of Pepsin and Trypsin," *Biological Reviews* 10 (1935): 263–282.

11. Letter from Seymour Cohen to Wendell Stanley, 24 January 1948, Wendell Stanley Papers. Carton 7, folder 71, Bancroft Library, University of California, Berkeley.

12. Schoenheimer was noted for his introduction of deuterium, the stable isotope of hydrogen, as a tracer in metabolic studies in collaboration with radiochemist Harold Urey and fellow biochemist David Rittenberg in the 1930s.

13. Lloyd M. Kozloff, "Biochemical Studies of Virus Reproduction" (Ph.D. dissertation, University of Chicago, 1949), 1, 69.

14. J. Howard Brown, "The Biological Approach to Bacteriology," *Journal of Bacteriology* 23, no. 1 (1932): 1–10.

15. Muller, "Physics and Fundamental Problems," 210–214.

6

Nucleation

1. Robert Gregg Frank, *Harvey and the Oxford Physiologists: A Study of Scientific Ideas* (Berkeley: University of California Press, 1980); Gerald L. Geison, *Michael Foster and the Cambridge School of Physiology: The Scientific Enterprise in Late Victorian Society* (Princeton: Princeton University Press, 1978); C. B. van Niel, "The 'Delft School' and the Rise of General Microbiology," *Bacteriological Reviews* 13(3) (1949): 161–174; Derek J. de Solla Price and Donald Beaver, "Collaboration in an Invisible College," *American Psychologist* 21, no. 11 (1966): 1011–1018; also Jerold W. Grossman, "Paul Erdős: The Master of Collaboration," *Algorithms and Combinatorics* 14 (1997): 467–475.

2. Mullins, "The Development of a Scientific Specialty," 51–82; Edwin Boring, *A History of Experimental Psychology*, 2nd ed. (Englewood Cliffs, N.J.: Prentice-Hall, 1950); Barbara Von Eckardt, *What Is Cognitive Science?* (Cambridge, Mass.: MIT Press, 1993).

3. Reginald Haydn Hopkins, "Some Recent Advances in the Biology of Malting and Brewing: The Artificial Hybridisation of Yeasts," *Journal of the Institute of Brewing* 46 (1940): 68–74; Frederick David Richey, *Corn Breeding* No. 1489. U.S. Department of Agriculture, 1927.

4. Hermann Joseph Muller, "Variation Due to Change in the Individual Gene," *The American Naturalist* 56, no. 642 (1922): 48–49.

5. Thomas D. Brock, "The Bacterial Nucleus: A History," *Microbiological Reviews* 52, no. 4 (1988): 397–411.

6. Harrison R. Stewart, "Radiation Therapy," *Engineering and Science* 32, no. 9 (1969): 14–16. Caltech website accessed 25 May 2021, https://resolver.caltech.edu/CaltechES:32.9.harrison.

7. Emory L. Ellis Papers, 1925–1993, 10126-MS, California Institute of Technology Archives and Special Collections, Online Guide: https://oac.cdlib.org/findaid/ark:/13030/tf167n97q6/; Kevin Struhl, "From *E. coli* to Elephants," *Nature* 417, no. 6884 (2002): 22–23.

8. William B. Coley, "The Treatment of Inoperable Sarcoma by Bacterial Toxins (The Mixed Toxins of the *Streptococcus erysipelas* and the *Bacillus prodigiosus*)," *Proceedings of the Royal Society of Medicine* 3, no. Surg_Sect (1910): 1–48. See also Alexander Brunschwig, "The Efficacy of 'Coley's Toxin' in the Treatment of Sarcoma: An Experimental Study," *Annals of Surgery* 109, no. 1 (1939): 109–113.

9. Fischer and Lipson, *Thinking About Science*, 143.

10. Fischer and Lipson, *Thinking About Science*, 57, 60.

11. Phillip R. Sloan and Brandon Fogel, eds., *Creating a Physical Biology: The Three-Man Paper and Early Molecular Biology* (Chicago: University of

Chicago Press, 2011); Erwin Schrödinger, *What Is Life?: The Physical Aspect of the Living Cell; Based on Lectures Delivered Under the Auspices of the Institute at Trinity College, Dublin, in February 1943* (Cambridge: Cambridge University Press, 1944).

12. Letter from Emory Ellis to author, 26 September 1991, in author's files.

13. Max Delbrück, Oral History Transcript, ID 1979-00005, California Institute of Technology Archives and Special Collections, 64.

14. Emory L. Ellis Papers, Interview with author, 5 March 1992, ID 10126-MS. California Institute of Technology Archives and Special Collections.

15. Letter from Salvador Luria to Emory Ellis, 2 October 1940, Emory L. Ellis Papers, folder 1.2, California Institute of Technology Archives and Special Collections; see also Luria, *A Slot Machine*, 31.

16. Thomas H. Morgan to the Rockefeller Foundation, June 1939, quoted in Fischer and Lipson, *Thinking About Science*, 126–127; Luria, *A Slot Machine*, 35.

17. Morgan, quoted in Fischer and Lipson, *Thinking About Science*, 126–127; see also Anonymous, "Silent Partners," *Vanderbilt Magazine*, 11 March 2008.

18. Luria, *A Slot Machine*, 37. Bloomington and Nashville are about 250 miles apart, not an easy journey in 1943 because of the absence of major highways and, more important, the imposition of gasoline rationing during World War II.

19. Quotation in Fischer and Lipson, *Thinking About Science*, 148, 140.

20. Alfred Hershey, *The Max Delbrück Laboratory Dedication Ceremony* (Cold Spring Harbor, N.Y.: Cold Spring Harbor Laboratory, 1981), 22.

21. Luria, *A Slot Machine*, 42; Fischer and Lipson, *Thinking About Science*, 149; Hershey, *Delbrück Laboratory Dedication*, 22.

22. Salvador E. Luria and Max Delbrück, "Mutations of Bacteria from Virus Sensitivity to Virus Resistance," *Genetics* 28, no. 6 (1943): 491–511.

23. Luria and Delbrück, "Mutations of Bacteria from Virus Sensitivity."

24. Alistair C. R. Dean and Cyril Hinshelwood, "Observations on Bacterial Adaptation," *Symposia of the Society for General Microbiology (London)* 3 (1953): 21–39. Joshua Lederberg and Esther M. Lederberg, "Replica Plating and Indirect Selection of Bacterial Mutants," *Journal of Bacteriology* 63, no. 3 (1952): 399–406.

25. Patrick F. Daegelen, William Studier, Richard E. Lenski, Susan Cure, and Jihyun F. Kim, "Tracing Ancestors and Relatives of *Escherichia coli* B, and the Derivation of B Strains REL606 and BL21 (DE3)," *Journal of Molecular Biology* 394, no. 4 (2009): 634–643. Demerec and Fano, "Bacteriophage-resistant Mutants." The New York Post-Graduate Medical School and Hospital was founded in 1882 and was completely absorbed into New York

University in 1946. Norton D. Zinder, ed., *RNA Phages* (Cold Spring Harbor, N.Y.: Cold Spring Harbor Laboratory Press, 1975); David T. Denhardt, David Dressler, Daniel S. Ray, eds., *The Single-Stranded DNA Phages* (Cold Spring Harbor, N.Y.: Cold Spring Harbor Laboratory Press, 1978); Alfred D. Hershey, ed., *The Bacteriophage Lambda* (Cold Spring Harbor, N.Y.: Cold Spring Harbor Laboratory Press, 1971).

26. Vannevar Bush, *Science—The Endless Frontier: A Report to the President by Vannevar Bush* (Washington: Director of the Office of Scientific Research and Development, United States Government Printing Office, 1945).

27. James Craigie and Chun Hui Yen, "The Demonstration of Types of *B. typhosus* by Means of Preparations of Type II Vi Phage: I. Principles and Technique," *Canadian Public Health Journal* 29, no. 9 (1938): 448–463.

28. John Howard Northrop, *Crystalline Enzymes: The Chemistry of Pepsin, Trypsin, and Bacteriophage* (New York: Columbia University Press, 1939). Ellis Papers, Interview with author, 5 March 1992. Earl Alison Evans, Jr., *Biochemical Studies of Bacterial Viruses* (Chicago: University of Chicago Press, 1952), 68.

29. Rasmussen, "Mid-Century Biophysics Bubble."

30. Cyrus Levinthal, "A Study of Protons Ejected from Nuclei by High Energy Gamma Rays" (Ph.D. dissertation, University of California, Berkeley, 1950). Robley C. Williams and Ralph W. G. Wyckoff, "The Thickness of Electron Microscopic Objects," *Journal of Applied Physics* 15, no. 10 (1944): 712–716. The first RCA model EMB was available in the fall of 1940. In the spring of 1941, O. S. Duffendack, an early proponent of electron microscopic studies in material science, obtained an RCA EMB for the department of physics at Michigan as soon as it was commercially available. Cyrus Levinthal and Harold W. Fisher, "The Structural Development of a Bacterial Virus," *Biochimica et Biophysica Acta* 9 (1952): 419–429. "We wish to thank Dr. M. Baylor, Dr. A. D. Hershey and Dr. S. E. Luria for suggesting improvements on an early draft of this paper, but even more for a great deal of help and advice on general problems of phage technique."

31. Aaron Novick and Leo Szilard, "Virus Strains of Identical Phenotype but Different Genotype," *Science* 113, no. 2924 (1951): 34–35; Aaron Novick and Leo Szilard, "Genetic Mechanisms in Bacteria and Bacterial Viruses I," *Cold Spring Harbor Symposium on Quantitative Biology* 16 (1951): 337–342; Aaron Novick and Leo Szilard, "Experiments on Light-Reactivation of Ultra-Violet Inactivated Bacteria," *Proceedings of the National Academy of Sciences of the United States of America* 35, no. 10 (1949): 591–600. Franck was director of the chemistry division of the Metallurgical Lab during World War II and winner of the 1925 Nobel Prize in Physics; he worked on photosynthesis. Urey was a geochemist and winner of the 1934 Nobel Prize in Chemistry;

along with his student Stanley Miller, Urey became famous for work on pre-
biotic organic synthesis.

32. Letter from Max Delbrück to Milislav Demerec, 21 May 1946. Max
Delbrück Papers 10045-MS, Box 28, folder 13, California Institute of Technology Archives and Special Collections, permission courtesy of the Delbrück family. Jean Weigle, "Story and Structure of the λ Transducing Phage,"
in Cairns, Stent, and Watson, *Phage and the Origins*, 227.

33. Max Delbrück Papers, ID 10045-MS, California Institute of Technology Archives and Special Collections. Delbrück's copies of the *Phage Information Service* show the mailing list in the margins, and one can follow recipients over time as well as see additions and deletions of recipients for each
issue of the newsletter. Letter from Gunther S. Stent to author, 30 December
1995.

34. Interviews with Seymour Cohen, Lloyd Kozloff, and Wacław Szybalski by the author in the mid-1990s.

35. Delbruck, Oral History Transcript, 73.

36. Ellis Papers, Interview with author, 5 March 1992; Frank W. Stahl, "Alfred Day Hershey, 1908–1997," *Biographical Memoirs of the National Academy of Sciences of the United States of America* 80 (2001): 142–159 (quotation
on p. 152).

7
Building the Group

1. Luria, *A Slot Machine*, 215–216.

2. Letter from Max Delbrück to Milislav Demerec, 6 November 1945,
Cold Spring Harbor Laboratory Archives, Demerec Files, permission courtesy of the Delbrück family.

3. Indiana University Biological Station, *Announcement of the Courses of
Instruction in the Spring Term, Summer Session, and Biological Station* (Bloomington, Ind.: Indiana University, 1900), 43–44.

4. Jane Maienschein, "The Marine Biological Laboratory Embryology
Course," *Embryo Project Encyclopedia* (2007-10-24), ISSN: 1940-5030, http://
embryo.asu.edu/handle/10776/1794; Susan Spath, "Van Niel's Course in General Microbiology," *ASM News* 70, no. 8 (2004): 359–363.

5. Millard Susman, "The Cold Spring Harbor Phage Course (1945–1970):
A 50th Anniversary Remembrance," *Genetics* 139, no. 3 (1995): 1101–1106.

6. Delbrück, Oral History Transcript, 69.

7. Sadly, none of the original participants are still alive, and investigation
of available archives yields little information beyond this point. Delbrück,

Oral History Transcript, 69. Phyllis Margaretten married meteorologist Lester Machta in 1947, and she would later become a nutritionist. Phyllis Margaretten Machta, "Effects of a Diet Low in Fat, Sugar and Salt on Response to Exercise Training by Atherosclerotic Patients" (Ph.D. dissertation, University of Maryland, 1979). She seemed to be "one of the boys" as a fellow graduate student with Gunther Stent at the University of Illinois in the mid-1940s. See also Jacob Christian Jacobsen and Thorbjörn Sigurgeirsson, *The Decay Constant of RaC'* (Copenhagen: Commissioner, Munksgaard, 1943); see also University of Iceland, Institute of Earth Sciences website, accessed 25 May 2021 https://earthice.hi.is/thorbjorn_sigurgeirssons_aeromagnetic _maps_iceland. The American-Soviet Medical Society was founded in New York City in 1943. Its major objective was to keep American physicians informed of Soviet medical advances and to improve relations between the United States and the USSR. Mudd served as president in the organization's final years. As a result of the Cold War, it was disbanded in 1947.

8. Stent, *Nazis, Women, and Molecular Biology*, 310–311. Seymour Benzer, "Notes on Phage Course, June 28–July 17, 1948," copy supplied to author. Letter from Noel R. Rose to author, 28 December 1995, letter in author's files.

9. On the Enzyme Club, see Philip Siekevitz, "The Continuing Life of the Enzyme Club of New York City: The Growth of American Biochemistry from 1942 to 1982," *Transactions of the New York Academy of Sciences* 41, no. 1, Series II (1983): 213–232. Karl Gordon Lark to author, 22 December 1995, letter in author's files. Max Delbrück, "Preface," in Mark Adams, *Bacteriophages* (New York: Interscience Publishers, 1959).

10. Letter from Joseph S. Gots to author, 29 December 1995, and letter from Morris Schaeffer to author, 3 January 1996; letters in author's files.

11. Letter from Philip S. Owen, executive secretary of National Research Council to M. Demerec, 21 March 1946, Wendell Stanley Papers, Carton 20, folder 30, Bancroft Library, University of California, Berkeley. The 1946 course roster from the Cold Spring Harbor Laboratory Archives: Earl A. Evans, Jr. (University of Chicago), Birgit Vennesland (University of Chicago), David Perkins (Columbia), Elizabeth M. Miller (Rockefeller Institute), Vernon Bryson (Rutgers University), Howard B. Newcombe (Cold Spring Harbor Laboratory), Hans Gaffron (University of Chicago), Harriet Taylor (Rockefeller Institute), Seymour Cohen (Rockefeller Institute), Catherine Fowler (University of Pennsylvania), Mark Adams (New York University), and Roman J. Kutsky (Princeton University). The 1947 list: Annnabel(?) Avery, Geoffrey H. Beale, M. Grieg, Halldór Grimsson, Albert Kelner, Philip Morrison, Aaron Novick, Gerald Oster, Richard B. Roberts, Harvey D. Rothberg, Jr., Albert Schatz, Leo Szilard, Wolf Vishniac, and M. Willis.

12. Renato Dulbecco and Marguerite Vogt, "Plaque Formation and Isola-

tion of Pure Lines with Poliomyelitis Viruses," *Journal of Experimental Medicine* 99, no. 2 (1954): 167–182; Wolfgang Joklik, Gwendolyn M. Woodroofe, Ian H. Holmes, and Frank Fenner, "The Reactivation of Poxviruses: I. Demonstration of the Phenomenon and Techniques of Assay," *Virology* 11, no. 1 (1960): 168–184.

13. See, for example, John M. Ziman, *Public Knowledge: An Essay Concerning the Social Dimension of Science* (Cambridge: Cambridge University Press, 1968); Lorraine Daston, "The Ideal and Reality of the Republic of Letters in the Enlightenment," *Science in Context* 4, no. 2 (1991): 367–386; de Solla Price and Beaver, "An Invisible College." See also David L. Hull, *Science as a Process: An Evolutionary Account of the Social and Conceptual Development of Science* (Chicago: University of Chicago Press, 2010).

14. Mimeographs were common at the time for producing small print runs, using a low-cost duplicating machine that works by forcing ink through a stencil onto paper. By the late 1960s, mimeographs were gradually displaced by photocopying.

15. Lloyd Kozloff, interview by author in 2003. Kozloff was a member of the Chicago group led by Earl Evans, Jr.

16. Ton van Helvoort, "The Controversy Between John H. Northrop and Max Delbrück on the Formation of Bacteriophage: Bacterial Synthesis or Autonomous Multiplication?" *Annals of Science* 49, no. 6 (1992): 545–575.

17. Max Delbrück, ed., *Viruses 1950* (Pasadena, Calif.: Division of Biology of the California Institute of Technology, 1950).

18. Letter from Vernon Bryson to Gunther Stent, 20 October 1960. Gunther Stent Papers, Carton 2, folder 51, Bancroft Library, University of California, Berkeley.

8
Place in Science

1. Latarjet arrived in October 1945. In March 1946 Demerec wrote to request an extension of Latarjet's visit for at least six more months; 6 March 1946 letter, Cold Spring Harbor Laboratory Archives, Carnegie files, Latarjet folder. Quotation from letter from Raymond Latarjet to Milislav Demerec, 9 December 1946, Cold Spring Harbor Laboratory Archives, Carnegie files, Latarjet folder.

2. James D. Watson, "Growing Up in the Phage Group," in Cairns, Stent, and Watson, *Phage and the Origins*, 241; Luria, *A Slot Machine*, 35.

3. Stent, *Nazis, Women, and Molecular Biology*, 305.

4. Cairns, Stent, and Watson, preface to *Phage and the Origins*, ix–x.

5. Allan Campbell, "Phage Integration and Chromosome Structure. A Personal History," *Annual Review of Genetics* 41 (2007): 1–11.

6. Cairns, Stent, and Watson, preface to *Phage and the Origins*, ix.

7. Mary Jo Nye, "National Styles? French and English Chemistry in the Nineteenth and Early Twentieth Centuries," *Osiris* 8 (1993): 30–49; Jane Maienschein, "Epistemic Styles in German and American Embryology," *Science in Context* 4, no. 2 (1991): 407–427; Pnina G. Abir-Am, "The First American and French Commemorations in Molecular Biology: From Collective Memory to Comparative History," *Osiris* 14 (1999): 324–372. Geison, *Michael Foster*. Londa L. Schiebinger, *Nature's Body: Gender in the Making of Modern Science* (New Brunswick, N.J.: Rutgers University Press, 1993). Bruno Latour and Steve Woolgar, *Laboratory Life: The Construction of Scientific Facts* (Princeton: Princeton University Press, 2013); William Coleman and Frederic Lawrence Holmes, eds., *The Investigative Enterprise: Experimental Physiology in Nineteenth-Century Medicine* (Berkeley: University of California Press, 1988); Gerald L. Geison, "Scientific Change, Emerging Specialties, and Research Schools," *History of Science* 19, no. 1 (1981): 20–40.

8. An extensive account of the nineteenth-century growth of the marine biology research station is given in the entry on "Laboratory" in the *New International Encyclopedia*, 1905 edition, by James C. Lough, Robert William Hall, Alpheus Spring Packard, John Merle Coulter, Edward Bradford Titchener, and Herbert Treadwell Wade.

9. The Brooklyn Institute of Arts and Sciences evolved from the Brooklyn Apprentices' Library Association formed in 1824; it later gave rise to the Cold Spring Harbor Laboratory, the Brooklyn Museum, the Brooklyn Botanic Garden, and the Brooklyn Academy of Music. Jan Witkowski, *The Road to Discovery: A Short History of Cold Spring Harbor Laboratory* (Cold Spring Harbor, N.Y.: Cold Spring Harbor Laboratory Press, 2016).

10. Hugo Fricke and Sterne Morse, "The Action of X-rays on Ferrous Sulphate Solutions," *Philosophical Magazine* 7, no. 7 (1929): 129–141; Hugo Fricke and Edwin J. Hart, "The Oxidation of $Fe++$ to $Fe+++$ by the Irradiation with X-Rays of Solutions of Ferrous Sulfate in Sulfuric Acid," *Journal of Chemical Physics* 3, no. 1 (1935): 60–61. The latter publication was from the "Walter B. James Laboratory for Biophysics, The Biological Laboratory, Cold Spring Harbor, Long Island, New York."

11. Milislav Demerec, "Heredity and Radiation, 1," *Radiology* 27, no. 2 (1936): 217–229. On gene conferences see Wendell Stanley Papers, Carton 7, folder 150, Bancroft Library, University of California, Berkeley. Claude E. Shannon, "An Algebra for Theoretical Genetics" (Ph.D. dissertation, Massachusetts Institute of Technology, 1940). Alexander Hollaender, Eva R. Sansome, Esther Zimmer, and Milislav Demerec, "Quantitative Irradiation

Experiments with *Neurospora crassa*. II. Ultraviolet Irradiation," *American Journal of Botany* 32, no. 4 (1945): 226–235. Joshua Lederberg and Esther M. Lederberg, "Replica Plating and Indirect Selection of Bacterial Mutants," *Journal of Bacteriology* 63, no. 3 (1952): 399–406.

12. Letter from Merle A. Tuve to Wendell Stanley, 27 September 1946, Wendell Stanley Papers, Carton 10, folder 96, Bancroft Library, University of California, Berkeley.

13. Max Delbrück, Oral History Transcript, 61.

14. Luria, *A Slot Machine*, 32–35; Fischer and Lipson, *Thinking About Science*, 130.

15. Life at Cold Spring Harbor Laboratory in the 1950s was physically represented on the desk of my Ph.D. adviser, Wacław Szybalski, in the form of a small child's sailboat mounted in a stand for all to see. Across the stern was painted the name of the boat, "All Mine." After I had completed my degree, I asked the story of the boat. It seems that Szybalski and Luria had jointly owned a small sailboat for weekend outings but were frequently in competition for it. When Luria finally left the lab and sold his interest to Szybalski, Luria gave him the toy boat as a reminder of their shared enterprise. On a visit to Hershey's lab in 1966, I was told that Hershey would appear after his afternoon nap (said to ward off his migraines), at which time he would wash up the day's glassware while we chatted.

16. The Carnegie Institution of Washington, which supported genetic research at Cold Spring Harbor Laboratory, steadfastly opposed the efforts by the staff scientists to seek additional outside research funding, on the grounds that such support, especially from the federal government, would diminish the "elite status" of the laboratory. Hershey, in need of more support, finally contravened the Carnegie board of directors and started to apply for federal grants in 1955. See Franklin W. Stahl, ed., *We Can Sleep Later: Alfred D. Hershey and the Origins of Molecular Biology* (Cold Spring Harbor, N.Y.: Cold Spring Harbor Laboratory Press, 2000), 29, 103.

17. Most of the scientific staff lived in rental housing on the grounds owned by the laboratory. A few lived in their own houses on property adjacent to the lab or nearby. In 1954 the costs for summer visitors at the laboratory were as follows: for a laboratory room, with the usual equipment, $100–$200 for the summer, depending on which building was used; an apartment with two bedrooms, living room, kitchen, and bath, $310 for the season; a two-room-and-bath bungalow without kitchen, $200, and with an extra room, $250. Dining room charges were $17 per week, per person, and half-price for children under five. Letter from Milislav Demerec to Seymour Benzer, 5 March 1954, Seymour Benzer Papers, ID 10242-MS, Carton 3, folder 12, California Institute of Technology Archives and Special Collections.

18. Letter from Milislav Demerec to Benzer, Doermann, Levinthal, Stent, and Watson, 17 November 1954. Seymour Benzer Papers, Carton 3, folder 12, California Institute of Technology Archives and Special Collections. Letter from Milislav Demerec to Benzer, Stent, and Watson, 21 February 1955. Gunther Stent Papers, Carton 3, folder 30, Bancroft Library, University of California, Berkeley.

9
The Challenge of Lysogeny

1. Félix d'Herelle, "La Théorie de l'aulolyse microbienne transmissible de Bordet et Ciucă," *Comptes rendus des séances de la Société de biologie et de ses filiales* 93 (1925): 1206–1208. Laurent Loison, Jean Gayon, and Richard M. Burian, "The Contributions—and Collapse—of Lamarckian Heredity in Pasteurian Molecular Biology: 1. Lysogeny, 1900–1960," *Journal of the History of Biology* 50, no. 1 (2017): 5–52.

2. Jules Bordet and Mihai Ciucă, "Exsudats leucocytaires et autolyse microbienne transmissible," *Comptes rendus des séances de la Société de biologie et de ses filiales* 83 (1920): 1293–1295. M. Sefer and M. Sefer, "La 10 ani de la moartea profesorului Mihai Ciucă (18. VII. 1883–20. II. 1969) Mihai Ciucă şi studiul bacteriofagului, însemnări în caietele de lucru" [10 years after the death of Professor Mihai Ciucă (18 July 1883–20 February 1969) Mihai Ciucă and the study of bacteriophage, recorded in his working notebooks], *Revista de Igiena, Bacteriologia, Virusologia, Parazitologia, Epidemiologia* 24 (2) (1979): 123–126. Marcel Lisbonne and Louis Carrère, "Sur l'obtention du principe bactériophagique par antagonisme microbienne," *Comptes rendus des séances de la Société de biologie et de ses filiales* 87 (1922): 1011.

3. Earl B. McKinley and Julia Cámara, "Bacteria as 'Carriers' of Bacteriophage," *Proceedings of the Society for Experimental Biology and Medicine* 27, no. 8 (1930): 847–848.

4. Eugène Wollman, "Recherches sur la bactériophagie (phénomène de Twort-d'Hérelle)," *Annales de l'Institut Pasteur* 39 (1925): 789–832.

5. Thomas D. Brock, "The Bacterial Nucleus: A History," *Microbiological Reviews* 52, no. 4 (1988): 397–411.

6. Neeraja Sankaran, "Mutant Bacteriophages, Frank Macfarlane Burnet, and the Changing Nature of 'Genespeak' in the 1930s," *Journal of the History of Biology* 43, no. 3 (2010): 571–599.

7. John H. Northrop, "Increase in Bacteriophage and Gelatinase Concentration in Cultures of *Bacillus megatherium*," *Journal of General Physiology* 23 (1939): 59–79.

8. William Whiteman Carlton Topley and Graham Selby Wilson, *The Principles of Bacteriology and Immunity*, 2nd ed. (London: Edward Arnold, 1936): 1417.

9. Max Delbrück, "Bacterial Viruses (Bacteriophages)," *Advances in Enzymology* 2 (1942): 4–5.

10. Max Delbrück, "Bacterial Viruses or Bacteriophage," *Biological Reviews* 21 (1946): 33.

11. Alfred D. Hershey and Jacob Bronfenbrenner, "Bacterial Viruses: Bacteriophages," in *Viral and Rickettsial Diseases of Man*, ed. Thomas M. Rivers (Philadelphia: J. B. Lippincott, 1948), 153–154.

12. J. S. K. Boyd, "The Dysenteries," *British Medical Journal* 1, no. 4720 (1951): 1440.

13. J. S. K. Boyd, "The Symbiotic Bacteriophages of *Salmonella typhimurium*," *Journal of Pathology and Bacteriology* 62.4 (1950): 501–517; J. S. K. Boyd and Benjamin Portnoy, "Bacteriophage Therapy in Bacillary Dysentery," *Transactions of the Royal Society of Tropical Medicine and Hygiene* 37, no. 4 (1944): 243–262. Also see J. S. K. Boyd, "Bacteriophage," *Biological Reviews* 31, no. 1 (1956): 72.

14. William C. Summers, "From Culture as Organism to Organism as Cell: Historical Origins of Bacterial Genetics," *Journal of the History of Biology* 24 (1991): 171–190.

15. On the secretion hypothesis see Donald A. Marvin and Barbara Hohn, "Filamentous Bacterial Viruses," *Bacteriological Reviews* 33, no. 2 (1969): 172–209. For lysogenic studies see André Lwoff and Antoinette Gutmann, "Recherches sur un *Bacillus megatherium* lysogène," *Annales de l'Institut Pasteur* 78, no. 6 (1950): 711–739.

16. Max E. Gottesman and Robert A. Weisberg, "Little Lambda, Who Made Thee?" *Microbiology and Molecular Biology Reviews* 68, no. 4 (2004): 796–813. Giuseppe Bertani, "Lysogeny at Mid-Twentieth Century: P1, P2 and Other Experimental Systems," *Journal of Bacteriology* 186 (2004): 595–600.

17. Richard M. Burian and Jean Gayon, "Un évolutionniste Bernardien à l'Institut Pasteur? Morphologie des ciliés et évolution physiologique dans l'œuvre d'André Lwoff," *L'Institut Pasteur: Contribution à son histoire* (Paris: Ed. de la Découverte, 1991), 165–186; also: Nadine Peyrieras and Michel Morange, "The Study of Lysogeny at the Pasteur Institute (1950–1960): An Epistemologically Open System," *Studies in History and Philosophy of Science Part C: Studies in History and Philosophy of Biological and Biomedical Sciences* 33, no. 3 (2002): 419–430.

18. William C. Summers, "Plasmids: Histories of a Concept," in *Reticulate Evolution: Symbiogenesis, Lateral Gene Transfer, Hybridization, and Infectious*

Heredity, ed. Nathalie Gontier (Cham, Switzerland: Springer Nature, 2015), 179–190. Allan M. Campbell, "Episomes," *Advances in Genetics* 11 (1963): 101–145; Allan Campbell, "Phage Integration and Chromosome Structure. A Personal History," *Annual Review of Genetics* 41 (2007): 1–11.

19. Joshua Lederberg, Esther M. Lederberg, Norton D. Zinder, and Ethelyn R. Lively, "Recombination Analysis of Bacterial Heredity," *Cold Spring Harbor Symposia on Quantitative Biology* 16 (1951): 413–443; Norton D. Zinder and Joshua Lederberg, "Genetic Exchange in *Salmonella,*" *Journal of Bacteriology* 64, no. 5 (1952): 679–699. Oswald T. Avery, Colin M. McLeod, and Maclyn McCarty, "Studies on the Chemical Nature of the Substance Inducing Transformation of Pneumococcal Types," *Journal of Experimental Medicine* 79, no. 2 (1944): 37–158. Milislav Demerec and Philip E. Hartman, "Complex Loci in Microorganisms," *Annual Reviews in Microbiology* 13, no. 1 (1959): 377–406.

20. M. Lee Morse, Esther M. Lederberg, and Joshua Lederberg, "Transduction in *Escherichia coli* K-12," *Genetics* 41, no. 1 (1956): 142–156.

21. Gontier, *Reticulate Evolution;* also Carlos Canchaya, Ghislain Fournous, Sandra Chibani-Chennoufi, Marie-Lise Dillmann, and Harald Brüssow, "Phage as Agents of Lateral Gene Transfer," *Current Opinion in Microbiology* 6, no. 4 (2003): 417–424.

22. Armin Dale Kaiser, "A Genetic Analysis of Bacteriophage Lambda" (Ph.D. dissertation, California Institute of Technology, 1955); Raymond K. Appleyard, "Segregation of Lambda Lysogenicity During Bacterial Recombination in *E. coli* K-12," *Cold Spring Harbor Symposia on Quantitative Biology* 18 (1953): 95–97; François Jacob and Elie L. Wollman, "Genetic Study of a Temperate Phage of *Bact. coli.* I. The Genetic System of Bacteriophage A," *Annales l'Institut Pasteur* 87, no. 6 (1954): 653–673.

23. William C. Summers, "From Enzyme Adaptation to Gene Regulation," *Advances in Applied Microbiology* 52 (2003): 159–166.

10

The Challenge of Phage Diversity

1. Robert L. Sinsheimer, "Multum in Parvo," in Cairns, Stent, and Watson, *Phage and the Origins,* 260. Vladimir Sertić and Nicolai Boulgakov, "Classification et identification des typhi-phages," *Comptes rendus des séances de la Société de biologie et de ses filiales* 119 (1935): 1270–1272. In this paper they used a classification based on size and serological relationships; the Greek letter φ represents phages with a wide host range, the serological group by X (ten) and the isolate number, 174. F. Macfarlane Burnet and Dora Lush,

"The Staphylococcal Bacteriophages," *Journal of Pathology and Bacteriology* 40, no. 3 (1935): 455–469. Chargaff's Rules describe the regularities in nucleotide base ratios observed in a wide variety of DNAs by Erwin Chargaff and his colleagues in the early 1950s. The molar ratios of adenine to thymine and guanine to cytosine were always 1.0, a key clue to base pairing for Watson and Crick.

2. Sinsheimer, "Multum in Parvo," 263.

3. D. E. Bradley, "Some New Small Bacteriophages (φX174 Type)," *Nature* 195, no. 4841 (1962): 622–623.

4. G. Nigel Godson, "Evolution of φX174. Isolation of Four New φX-like Phages and Comparison with φX174," *Virology* 58, no. 1 (1974): 272–289. Kasper Zechel, Jean-Pierre Bouché, and Arthur Kornberg, "Replication of Phage G4. A Novel and Simple System for the Initiation of Deoxyribonucleic Acid Synthesis," *Journal of Biological Chemistry* 250, no. 12 (1975): 4684–4689.

5. Hartmut Hoffmann-Berling, Donald A. Marvin, and Hildegard Dürwald, "Ein fädiger DNS-Phage (fd) und ein sphärischer RNS-Phage (fr), wirtsspezifisch für männliche Stämme von *E. coli*: 1. Präparation und chemische Eigenschaften von fd und fr," *Zeitschrift für Naturforschung B* 18, no. 11 (1963): 876–883; Peter Hans Hofschneider, "Untersuchungen über kleine *E. coli* K 12 Bakteriophagen," *Zeitschrift für Naturforschung B* 18, no. 3 (1963): 203–210.

6. Tim Loeb and Norton D. Zinder, "A Bacteriophage Containing RNA," *Proceedings of the National Academy of Sciences of the United States of America* 47, no. 3 (1961): 282–289.

7. William Paranchych and Angus F. Graham, "Isolation and Properties of an RNA-containing Bacteriophage," *Journal of Cellular and Comparative Physiology* 60, no. 3 (1962): 199–208; James E. Davis, Robert L. Sinsheimer, and James H. Strauss, "Bacteriophage MS2—Another RNA phage," *Science* 134, no. 348 (1961): 1427.

8. One aspect of the APG culture that was widely respected was the obligation to supply strains, mutant organisms, and other experimental material upon request, with the reciprocal obligation not to request such help in order to do experiments in competition with the supplier. R17 phage was distributed widely in response to the perceived unavailability of phage f2.

9. Itaru Watanabe, Tadashi Miyake, Toshizo Sakurai, Tadayoshi Shiba, and Tsuneya Ohno, "Isolation and Grouping of RNA Phages," *Proceedings of the Japan Academy* 43, no. 3 (1967): 204–209; Lacy R. Overby, Grant H. Barlow, Roy H. Doi, Monique Jacob, and Sol Spiegelman, "Comparison of Two Serologically Distinct Ribonucleic Acid Bacteriophages I. Properties of the Viral Particles," *Journal of Bacteriology* 91, no. 1 (1966): 442–448; Ichiro

Haruna and Sol Spiegelman, "Specific Template Requirements of RNA Replicases," *Proceedings of the National Academy of Sciences of the United States of America* 54, no. 2 (1965): 579–587.

11
The French Connection

1. Fleming and Bailyn, *The Intellectual Migration.*

2. Jean-Paul Gaudillière, "Paris–New York Roundtrip: Transatlantic Crossings and the Reconstruction of the Biological Sciences in Postwar France," *Studies in History and Philosophy of Science Part C: Studies in History and Philosophy of Biological and Biomedical Sciences* 33, no. 3 (2002): 389–417.

3. Doris T. Zallen, "Louis Rapkine and the Restoration of French Science After the Second World War," *French Historical Studies* (1991): 6–37.

4. Doris T. Zallen, "The Rockefeller Foundation and French Research," *Cahiers pour l'histoire du CNRS* 5, no. 24 (1989): 35–58.

5. François Jacob, Interview by Michel Morange, October 2004, Track 34, "The American Links," *Web of Stories* accessed 25 May 2021, www.webof stories.com/play/francois.jacob/1.

6. Maurice Tubiana, "Raymond Latarjet, A Scientist of the Century," *Comptes rendus des séances de la Société de biologie et de ses filiales* 192, no. 3 (1998): 383–385. As a person and as an intellect, Latarjet was open to many ideas and cultures. His wife was a professor of music at the Sorbonne, and he prided himself on his connections with the arts and humanities. He was particularly pleased to deliver a program on French radio on the subject of Schubert's song cycle "Die Winterreise." His interest in the avant-garde French poets and his appreciation of American baseball impressed me as well. Francis-André Wollman, "Elie Wollman, un homme de conviction. Une défense de la recherche incarnée par des femmes, des hommes et leurs institutions," *Histoire de la recherche contemporaine. La revue du Comité pour l'histoire du CNRS* 7, no. 2 (2018): 202–211. Michel Morange, "What History Tells Us, III. André Lwoff: From Protozoology to Molecular Definition of Viruses," *Journal of Biosciences* 30, no. 5 (2005): 591–594.

7. Angela N. H. Creager, *Life Atomic: A History of Radioisotopes in Science and Medicine* (Chicago: University of Chicago Press, 2013).

8. Latarjet interview with author, 20 August 1993. Latarjet's Parisian lab notebooks for work done prior to coming to America clearly showed results from preliminary "L-L" experiments.

9. Summers, "Enzyme Adaptation," 159–166. Lwoff quoted in Cairns, Stent, and Watson, *Phage and the Origins,* 89.

10. Cairns, Stent, and Watson, *Phage and the Origins,* 216–217.

11. Morange, "What History Tells Us," 591–594. John Harley Warner, "Remembering Paris: Memory and the American Disciples of French Medicine in the Nineteenth Century," *Bulletin of the History of Medicine* 55 (1991): 301–325.

12. Gaudillière, "Paris–New York Round Trip," 407. Margaret Lieb, "Poisson d'Avril," unpublished essay, copy supplied to the author, December 1995.

13. Jean-Pierre Gratia, "André Gratia: A Forerunner in Microbial and Viral Genetics," *Genetics* 156, no. 2 (2000): 471–476. Masayasu Nomura, "Colicins and Related Bacteriocins," *Annual Review of Microbiology* 21, no. 1 (1967): 257–284.

14. Kay, *Molecular Vision of Life.*

15. André Lwoff, "Introduction: Le Bactériophage: Premier colloque international 1952," *Annales de l'Institut Pasteur* 84, no. 1 (1953): 3–4.

16. Lwoff, "Le Bactériophage," 3–313.

17. Presenters named on the program but omitted from Delbrück's list are Seymour Cohen (U.S.), Gerard R. Wyatt (U.S.), Raymond Latarjet (France), and his own student, George H. Bowen (U.S.), although Delbrück may have delivered Bowen's paper in his stead. A list identifying people in a photograph of attendees included the names of nearly two dozen people not mentioned by Delbrück. Some were non-scientist spouses and some were scientists. Gunther Stent Papers, Carton 6, folder 4, Bancroft Library, University of California, Berkeley.

18. Max Delbrück, "Summary for the Grandees of the National Foundation for Infantile Paralysis," 16 pp. Copy in the Frank W. Putman Papers, Archives of the American Society of Microbiology. Although the American biochemist Seymour Cohen was not listed by Delbrück as being present at the Royaumont meeting, Cohen's protégé, Lawrence Weed, was present. Weed had just published several papers using radio-labeled precursors of nucleic acids to study just this point. Weed left phage research in the mid-1960s to become famous as the originator of the "problem-oriented medical record," the SOAP (subjective, objective, assessment, and plan) medical note; he has been hailed as the "father of medical informatics."

19. Sandra Citi and Douglas E. Berg, "Grete Kellenberger-Gujer: Molecular Biology Research Pioneer," *Bacteriophage* 6, no. 2 (2016): 1–12. Website accessed 25 May 2021, http://dx.doi.org/10.1080/21597081.2016.1173168. For an analysis of the uniqueness of the Pasteur Institute in French culture, see Bruno Latour, *The Pasteurization of France* (Cambridge: Harvard University Press, 1993).

20. Letter from Gunther Stent to I. J. Barnes, 6 February 1979. Gunther

Stent Papers, Carton 2, folder 47, Bancroft Library, University of California, Berkeley.

21. Max Delbrück and Robert Edgar, "Jean-Jacques Weigle, 1901–1968," *Engineering and Science* 32, no. 4 (1969): 21. The restriction of phage growth on one host strain depending on the host strain in which the phage had been prepared (i.e., modified) was a strange phenomenon discovered independently in 1952 by Weigle and Bertani in λ phage and Luria and Human in T2 and T6 phage. At the time it was called host-controlled-modification. The underlying mechanism is based on DNA sequence-specific endonucleases ("restriction enzymes"), which, as gene-spicing enzymes, form the basis for much of the current biotechnology industry.

22. Letter from Joe Bertani to Noreen Murray, 17 July 2003, reprinted in appendix 1, chapter 1, Wil A. M. Loenen, *Restriction Enzymes: A History* (Cold Spring Harbor, N.Y.: Cold Spring Harbor Laboratory Press, 2019). Kellenberger, "History of Phage Research," 15.

23. Sankaran, "Burnet and Bacteriophage." Also: Macfarlane Burnet, *Changing Patterns: An Atypical Autobiography* (Melbourne: William Heinemann, 1968).

24. Soraya De Chadarevian, *Designs for Life: Molecular Biology After World War II* (Cambridge: Cambridge University Press, 2002). Boyd and Portnoy, "Bacteriophage Therapy, 243–262; Boyd, "Symbiotic Bacteriophages," 501–517. William C. Summers, "Sydney Brenner (1927–2019)," *Annual Review of Virology* 6 (2019): i–ii. https://doi.org/10.1146/annurev-vi-06-072619-100111; Sydney Brenner and Gunther S. Stent, "Bacteriophage Growth in Protoplasts of *Bacillus megaterium*," *Biochimica et biophysica acta* 17 (1955): 473–475. Quotation from letter from Gunther Stent to F. T. Wall, 24 April 1958, Gunther Stent Papers, Carton 2, folder 36, Bancroft Library, University of California.

25. Paul G. Fildes, David Kay, and Wolfgang K. Joklik, "Divalent Metals in Phage Production," in *The Nature of Virus Multiplication: Second Symposium of the Society for General Microbiology Held at Oxford University, April 1952*, ed. Paul G. Fildes and William E. Van Heyningen (Cambridge: Cambridge University Press, 1953). Francis H. C. Crick, Leslie Barnett, Sydney Brenner, and Richard J. Watts-Tobin, "General Nature of the Genetic Code for Proteins," *Nature* 192, no. 4809 (1961): 1227–1232.

26. Kellenberger, "History of Phage Research." Notable young postdoctoral trainees in molecular biology moving from the United States to Geneva in this period include William Wood, Peter Moore, Richard Burgess, Edward N. Brody, Jerry Adams, Suzanne Cory, and Ann Baker Burgess. On postwar isotopes, see Creager, *Life Atomic*.

27. Nicolas Rasmussen, *Picture Control: The Electron Microscope and the Transformation of Biology in America, 1940–1960* (Stanford: Stanford University Press, 1999).

12
Laboratory Life

1. Before delivering my first seminar at Caltech I was warned that Delbrück would wait until I had a slide projected on the screen to get up and walk out, with his shadow growing larger and larger on the screen as he walked in the beam toward the projector. As the seminar progressed, I anxiously awaited this exit that never came, but I was grilled as expected in private discussion later.

2. Cairns, Stent, and Watson, *Phage and the Origins*, 241.

3. Fischer and Lipson, *Thinking About Science*, 38.

4. Ziman, *Public Knowledge*. Delbrück interview with the author, 1970. A comparison of the dates of each of these meetings, however, shows that such a conflict would have happened only a few times, so this was probably an aspiration in Delbrück's mind rather than a fact.

5. David K. Johnson, *The Lavender Scare: The Cold War Persecution of Gays and Lesbians in the Federal Government* (Chicago: University of Chicago Press, 2009).

6. Margaret W. Rossiter, *Women Scientists in America: Before Affirmative Action, 1940–1972*, vol. 2 (Baltimore: Johns Hopkins University Press, 1998); Pnina G. Abir-Am and Dorinda Ed Outram, *Uneasy Careers and Intimate Lives: Women in Science 1789–1979* (New Brunswick, N.J.: Rutgers University Press, 1987); Londa Schiebinger, *The Mind Has No Sex?: Women in the Origins of Modern Science* (Boston: Harvard University Press, 1991); Sandra G. Harding, *The Science Question in Feminism* (Ithaca, N.Y.: Cornell University Press, 1986); Bert Hansen, "Public Careers and Private Sexuality: Some Gay and Lesbian Lives in the History of Medicine and Public Health," *American Journal of Public Health* 92, no. 1 (2002): 36–44.

7. Even as late as the mid-1960s, there were significant numbers of unpaid faculty at major universities, people of independent means ("dollar-a-year men"). In the early 1970s, several young, recently hired faculty at my institution approached their department chair to inquire about their salaries; the response was, "You actually try to live on your salary?" Anti-nepotism policies were another barrier to equal treatment of married scientists in many academic institutions.

8. Luria, *A Slot Machine*, 43–44.

9. Letter from Max Delbrück to Milislav Demerec, 21 May 1946. Delbrück had already accepted three applicants from the University of Chicago (Earl Evans, Bridget Vennesland, and Hans Gaffron), and he evaluated the remaining applications for the 1946 phage course and recommended several as "Definitely wanted" (Seymour Cohen, Mark Adams, Howard Newcombe, and Vernon Bryson), "Doubtful candidates" (David Perkins, Elizabeth Miller, Harriet Taylor, ? Stumpf, Gao Zhangin), and "Definitely to be discouraged" (M. Wachstein, ? Scheierson, S. Stanley). Added to the last group was Delbrück's assessment, "I don't trust anybody connected with Mount Sinai." Max Delbrück Papers 10045-MS, Box 28, folder 13, California Institute of Technology Archives and Special Collections, permission courtesy of the Delbrück family.

10. Georgina Ferry, "History: Women in Crystallography," *Nature News* 505, no. 7485 (2014): 609. Foremost was Kathleen Lonsdale, first woman elected to the Royal Society of London in 1945. Others include Florence Bell, Dorothy Hodgkin, Rosalind Franklin, Barbara Low, and Isabella Karle. See also Julia Sanz-Aparicio, "The Legacy of Women to Crystallography," *ARBOR Ciencia, Pensamiento y Cultura* 191, no. 772 (2015): a216.

11. Fischer and Lipson, *Thinking About Science,* 137. Email from Jon Machta (son of Phyllis Margaretten) to author, 20 February 2019, in author's files. Stent to Treumann, 12 May 1948, Stent Papers.

12. Lieb, "Poisson d'Avril."

13. Luria, *A Slot Machine,* 210–215.

14. Hilary Rose and Steven Rose, "Red Scientist: Two Strands from a Life in Three Colors," in *J. D. Bernal: A Life in Science and Politics,* ed. Francis Aprahamian and Brenda Swann (London: Verso, 1999), 134.

15. Shirley M. Tilghman, "Science vs. Women—A Radical Solution," *New York Times,* 26 January 1993, Section A, page 23.

16. Barney G. Glaser, "Comparative Failure in Science," *Science* 143, no. 3610 (1964): 1012–1014. When I first read this paper in 1964, as a graduate student, I realized how helpful this perspective could be to an ambitious young scientist. Its lessons have remained with me for my entire professional career.

17. Ann Marie Skalka, "Finding, Conducting, and Nurturing Science: A Virologist's Memoir," *Annual Review of Virology* 4 (2017): 1–35.

18. Fischer and Lipson, *Thinking About Science,* 181–182. Cairns, Stent, and Watson, *Phage and the Origins,* 159.

19. Bert Hansen, "Has the Laboratory Been a Closet? Gay and Lesbian Lives in the History of Science," unpublished lecture at the National Library of Medicine, 15 June 2000, cited with the author's permission.

20. Milton W. Taylor, "Stanford University, Palo Alto, California (1961–66)"

IU Scholarworks. MW Taylor: Autobiography, Website accessed 25 May 2021, https://scholarworks.iu.edu/dspace/handle/2022/21716?show=full.

21. Obituary of Martha Chase, *New York Times,* 13 August 2013. Author interview with Thomas Pollard, 15 December 2020.

22. James D. Watson, *Genes, Girls, and Gamow* (New York: Knopf, 2002). Thomas A. Steitz to the author. Stent, *Nazis, Women, and Molecular Biology.*

23. When I was a sabbatical visitor at Cold Spring Harbor Laboratory in 1975 I observed an example of this performativity of informality while talking to Jim Watson, the director of the lab at the time. Watson mentioned that he was on his way to a meeting of the board of directors of the laboratory. He reached down, untied his shoes, and then rumpled his hair. He said they expected him to be eccentric and he did not want to disappoint them.

24. This attire was not usual for me, but I assumed that my colleagues would be the prep-school, Ivy League sort I had seen in movies and read about in books. Nothing could have been further from the truth, of course.

25. Kay, *Molecular Vision of Life.*

26. As the original members of the APG gave way to the next generation of scientists who identified more generally as molecular biologists, and as the leadership at Cold Spring Harbor Laboratory moved in various new directions, several of the emergent leaders of phage research and organizers of phage meetings were women.

27. Executive Order 10925, 6 March 1961, John F. Kennedy.

13

Maturation and Assimilation

1. Salvador E. Luria, "Reactivation of Ultraviolet-Irradiated Bacteriophage by Multiple Infection," *Journal of Cellular and Comparative Physiology* 39, no. Suppl 1 (1952): 119–123; Renato Dulbecco, "Reactivation of Ultra-Violet-Inactivated Bacteriophage by Visible Light," *Nature* 163, no. 4155 (1949): 949–950.

2. Thomas F. Anderson, "Morphological and Chemical Relations in Viruses and Bacteriophages," *Cold Spring Harbor Symposia on Quantitative Biology* 11 (1946): 1–13; also Thomas F. Anderson, "Electron Microscopy of Phages," in Cairns, Stent, and Watson, *Phage and the Origins,* 75.

3. Alfred Day Hershey, Martin D. Kamen, Joseph W. Kennedy, and Howard Gest, "The Mortality of Bacteriophage Containing Assimilated Radioactive Phosphorus," *Journal of General Physiology* 34, no. 3 (1951): 305–319; Gunther S. Stent and Clarence R. Fuerst, "Genetic and Physiological Effects

of the Decay of Incorporated Radioactive Phosphorus in Bacterial Viruses and Bacteria," *Advances in Biological and Medical Physics* 7 (1960): 1–75.

4. Watson, "X-Ray-Inactivated Bacteriophage I, II." See also Franklin Hutchinson, "Molecular Basis for Action of Ionizing Radiations," *Science* 134, no. 3478 (1961): 533–538.

5. Evelyn M. Witkin, "Genetics of Resistance to Radiation in *Escherichia coli*," *Genetics* 32, no. 3 (1947): 221–248; Ruth F. Hill, "A Radiation-Sensitive Mutant of *Escherichia coli*," *Biochimica et Biophysica Acta* 30, no. 3 (1958): 636–637. Stahl, "Alfred Day Hershey, 1908–1997," 145.

6. Richard P. Boyce and Paul Howard-Flanders, "Release of Ultraviolet Light-Induced Thymine Dimers from DNA in *E. coli* K-12," *Proceedings of the National Academy of Sciences of the United States of America* 51, no. 2 (1964): 293; Richard B. Setlow and William L. Carrier, "The Disappearance of Thymine Dimers from DNA: An Error-Correcting Mechanism," *Proceedings of the National Academy of Sciences of the United States of America* 51, no. 2 (1964): 226. Max Delbrück, "I was impatient with biochemistry in the sense of metabolic pathways converting one small molecule into another, and with the idea that further pursuit of this kind of biochemistry would lead to the nature of the gene, and its replication, and its effects." Oral History Transcript, 73.

7. Luria, "Reactivation." See, for examples, J. N. Davidson, *The Biochemistry of the Nucleic Acids* (New York: John Wiley & Sons, 1950); Van Rensselaer Potter, *Nucleic Acid Outlines*, vol 1. (Minneapolis: Burgess, 1960).

8. John Tyler Bonner, *First Signals: The Evolution of Multicellular Development* (Princeton: Princeton University Press, 2000).

9. Richard M. Burian, Jean Gayon, and Doris T. Zallen, "Boris Ephrussi and the Synthesis of Genetics and Embryology," in *A Conceptual History of Modern Embryology*, ed. Scott F. Gilbert (New York: Plenum, 1991), 207–227; Richard M. Burian and Jean Gayon, "The French School of Genetics: From Physiological and Population Genetics to Regulatory Molecular Genetics," *Annual Review of Genetics* 33, no. 1 (1999): 313–349; Richard M. Burian, Jean Gayon, and Doris Zallen, "The Singular Fate of Genetics in the History of French Biology, 1900–1940," *Journal of the History of Biology* 21, no. 3 (1988): 357–402. Max Neisser, "Ein Fall von Mutation nach de Vries bei Bakterien und andere Demonstrationen," *Zentralblatt für Bakteriologie, Parasitenkunde und Infektionskrankheiten I, Ref.*, 9f (1906): 98–102. Rudolf Massini, "Über einen in biologischer Bezeihung interessanten Kolistamm (*Bacterium coli mutabile*)," *Archiv für Hygiene* 61 (1907): 250–292.

10. E. Peter Geiduschek interview with author in 1970 about the recent discovery of the sigma factor that controls the action of *E. coli* RNA polymerase in copying the genes of phage T4.

11. As noted earlier, the belief that biological answers are "knowable" and that simplicity is preferable to complexity were two key principles of the APG. Vitalism and its relatives were still formidable in the decades when the APG flourished, found in the work of respected scientists such as Richard Goldschmidt, Ludwig von Bertalanffy, and James Lovelock, to name a few examples.

12. Letter from Sydney Brenner to Max Perutz, 5 June 1963, reprinted in *The Nematode Caenorhabditis elegans*, ed. William B. Wood et al. (Cold Spring Harbor, N.Y.: Cold Spring Harbor Laboratory, 1988).

13. Aldolase is an enzyme involved in the metabolism of sugar, which splits the sugar containing 6 carbon atoms to yield two molecules each with 3 carbon atoms. In spite of Brenner's apparent disdain, this reaction, and its reverse aldol condensation, is an amazingly important metabolic reaction that is used at many steps in the biochemistry of life to break (and make) chemical bonds between carbon atoms.

14
The American Phage Group as a
Model of Discipline Formation

1. Many historical accounts recognize a dual origin story: one based on genes and one based on macromolecular structure. See, for example, John C. Kendrew, "How Molecular Biology Started," *Scientific American* 216, no. 3 (1967): 141–144.

2. Michael S. Mahoney, "Computers and Mathematics: The Search for a Discipline of Computer Science," in *The Space of Mathematics: Philosophical, Epistemological, and Historical Explorations*, ed. Javier Echeverria, Andoni Ibarra, and Thomas Mormann (Berlin: Walter de Gruyter, 1992), 349–363.

3. Nadya T. Bliss, B. R. Erick Peirson, Deryc Painter, and Manfred D. Laubichler, "Anomalous Subgraph Detection in Publication Networks: Leveraging Truth," in *48th Asilomar Conference on Signals, Systems, and Computers*, ed. Michael B. Matthews (Piscataway, N.J.: Institute of Electrical and Electronics Engineers, 2014), 2005–2009. On discipline formation see, for example: Von Eckardt, *What Is Cognitive Science?*; Timothy Lenoir, *Instituting Science: The Cultural Production of Scientific Disciplines* (Stanford: Stanford University Press, 1997); Alexander Powell, Maureen A. O'Malley, Staffan Müller-Wille, Jane Calvert, and John Dupré, "Disciplinary Baptisms: A Comparison of the Naming Stories of Genetics, Molecular Biology, Genomics, and Systems Biology," *History and Philosophy of the Life Sciences* 29 (2007): 5–32; and Ellen Messer-Davidow, David R. Shumway, and David Sylvan, eds.,

Knowledges: Historical and Critical Studies in Disciplinarity (Charlottesville: University of Virginia Press, 1993). Gary Hatfield, "Wundt and Psychology as Science: Disciplinary Transformations," *Perspectives on Science* 5, no. 3 (1997): 349–382. Mullins, "The Development of a Scientific Specialty," 51–82. De Solla Price and Beaver, "An Invisible College."

4. Thomas Kuhn, *The Structure of Scientific Revolutions,* 2nd ed. Enlarged (Chicago: University of Chicago Press, 1970). Von Eckardt, *What Is Cognitive Science?* Edna Suarez-Diaz, "Molecular Evolution: Concepts and the Origin of Disciplines," *Studies in History and Philosophy of Science Part C: Studies in History and Philosophy of Biological and Biomedical Sciences* 40, no. 1 (2009): 50.

5. Emory L. Ellis and Max Delbrück, "The Growth of Bacteriophage," *Journal of General Physiology* 22, no. 3 (1939): 365–384.

6. Max Delbrück, "Problems of Modern Biology in Relation to Atomic Physics," 1944. A series of lectures given at the Vanderbilt University School of Medicine Library, April and May 1944. PB 60410, Second Printing (without change) February 1946 (University of Pennsylvania Library, MP539 D373, mimeographed typescript).

7. Delbrück, "Problems of Modern Biology."

8. Lily E. Kay, "Conceptual Models and Analytical Tools: The Biology of Physicist Max Delbrück," *Journal of the History of Biology* 18, no. 2 (1985): 207–246.

9. Schrödinger, *What Is Life and What Is Matter,* 49–61.

10. Muller, "Variation in the Individual Gene," 48–49.

11. Wacław Szybalski, "In Memoriam: Alfred D. Hershey (1908–1997)," in *We Can Sleep Later: Alfred D. Hershey and the Origins of Molecular Biology,* ed. Franklin W. Stahl (Cold Spring Harbor, N.Y.: Cold Spring Harbor Laboratory Press, 2000), 19. See also Frederic Lawrence Holmes, *Meselson, Stahl, and the Replication of DNA: A History of "The Most Beautiful Experiment in Biology"* (New Haven: Yale University Press, 2008).

12. François Jacob, David Perrin, Carmen Sánchez, and Jacques Monod, "L'opéron: groupe de gènes à expression coordonnée par un opérateur," *Comptes rendus hebdomadaires des séances de l'Academie des sciences* 250 (1960): 1727–1729.

13. This interpretation of concept transplantation was suggested by one of the anonymous reviewers of this manuscript, for which I am grateful.

Index